Selected Titles in This Series

D1283970

Poincaré and the Three Body Problem

DATE DUE

Dec 16	2006		

History of Mathematics
Volume 11

Poincaré and the Three Body Problem

June Barrow-Green

American Mathematical Society
London Mathematical Society

1991 *Mathematics Subject Classification.* Primary 01; Secondary 70.

Photographs on the cover are Henri Poincaré (inset) and Oscar II, King of Sweden and Norway (background).

A list of photograph and figure credits is included at the beginning of this volume.

Library of Congress Cataloging-in-Publication Data
Barrow-Green, June, 1953–
    Poincaré and the three body problem / June Barrow-Green.
        p. cm. — (History of mathematics, ISSN 0899-2428; v. 11)
    Includes bibliographical references (p.   –   ) and index.
    ISBN 0-8218-0367-0 (acid-free paper)
    1. Three-body problem. 2. Hamiltonian systems. 3. Poincaré, Henri, 1854–1912—Contributions in dynamics. I. Title. II. Series.
QA852.B37 1996
515′.352—dc20
                                                                                            96-11112
                                                                                                 CIP

For my Mother and Sister
and in memory of my Father and Brother
with love and thanks

Contents

Acknowledgements

This book derives from the PhD thesis I prepared at the Open University between 1989–1993, and I am very grateful to Jeremy Gray, who suggested the topic and patiently supervised the work involved. His help, enthusiasm, and kindness were unfailing and his scholarship an inspiration.

Several people both in this country and abroad have gone out of their way on my behalf and I extend thanks to them all. I particularly wish to thank the Institut Mittag-Leffler (shown below) for allowing me to use their archives, and whose staff provided me with every possible assistance during the time I spent there; Jesper Lützen, Roger Cooke, and Serguei Demidov for helping to make my visit to the Institut Mittag-Leffler so rewarding and enjoyable; and Steen Norgaard for providing insight into Scandinavian culture, and whose skills of translation and sense of humour were a constant source of delight.

I also wish to express thanks to my many friends and colleagues at the Open University who helped me in many and varied ways. In particular I wish to thank Derek Richards and Paul Dando for useful and stimulating discussions about Poincaré's dynamics; John Fauvel for sharing so generously his knowledge and expertise; and Marion Hall, Merrian Lancaster, and Angela Redgewell, who provided essential encouragement both on the tennis court and off.

Finally, immeasurable thanks go to Rory Collins, without whom my work would not have begun, and Ray Weedon, Chris Glen, and Marlese von Broembsen, without whom it would not have continued. Their support at different times and in different places meant more than words can say.

Photograph and Figure Credits

The AMS gratefully acknowledges the kindness of these individuals, institutions, and publishers in granting the following permissions.

June Barrow-Green

> Title page on cover and p. 242; from Henri Poincaré, *Sur le Problème des Trois Corps et les Équations de la Dynamique*, unpublished memoir; Courtesy of June Barrow-Green, private collection.

> Table of contents on pp. 243–244; from Henri Poincaré, *Sur le Problème des Trois Corps et les Équations de la Dynamique*, unpublished memoir; Courtesy of June Barrow-Green, private collection.

> Figure 6.i on cover and p. 249; from Henri Poincaré, *Sur le Problème des Trois Corps et les Équations de la Dynamique*, unpublished memoir, p. 43; Courtesy of June Barrow-Green, private collection.

Institut Mittag-Leffler

> Photograph of Jacques Hadamard on p. 200; from *Acta Mathematica* 1882–1912, *Table Générale des Tomes* 1–35, p. 144; Courtesy of the Institut Mittag-Leffler.

> Photograph of Charles Hermite on p. 54; from *Acta Mathematica* 1882–1912, *Table Générale des Tomes* 1–35, p. 145; Courtesy of the Institut Mittag-Leffler.

> Photograph of George Hill on p. 144; from *Acta Mathematica* 1882–1912, *Table Générale des Tomes* 1–35, p. 146; Courtesy of the Institut Mittag-Leffler.

> Photograph of Gösta Mittag-Leffler on p. 52; from *Acta Mathematica* 1882–1912, *Table Générale des Tomes* 1–35, p. 160; Courtesy of the Institut Mittag-Leffler.

> Photograph of Edvard Phragmén on p. 64; from *Acta Mathematica* 1882–1912, *Table Générale des Tomes* 1–35, p. 163; Courtesy of the Institut Mittag-Leffler.

> Photograph of Henri Poincaré on cover and p. xii; from *Acta Mathematica* 1882–1912, *Table Générale des Tomes* 1–35, p. 164; Courtesy of the Institut Mittag-Leffler.

Photograph of Karl Weierstrass on p. 56; from *Acta Mathematica* 1882–1912, *Table Générale des Tomes* 1–35, p. 176; Courtesy of the Institut Mittag-Leffler.

Letter from Poincaré to Mittag-Leffler on pp. 120-121; Courtesy of Institut Mittag-Leffler Archives.

Prix Oscar II on pp. 236–237; from Acta Mathematica **11** (1888), pp. 401–402; Courtesy of the Institut Mittag-Leffler.

Title page on p. 245; from Henri Poincaré, *Sur le Problème des Trois Corps et les Équations de la Dynamique*, Acta Mathematica, **13** (1890); Courtesy of the Institut Mittag-Leffler.

Table of Contents on pp. 246–247; from Henri Poincaré, *Sur le Problème des Trois Corps et les Équations de la Dynamique*, Acta Mathematica, **13** (1890); Courtesy of the Institut Mittag-Leffler.

Figures 5.5.i and 5.5.ii on p. 90; from Henri Poincaré, *Sur le Problème des Trois Corps et les Équations de la Dynamique*, Acta Mathematica, **13** (1890); Courtesy of the Institut Mittag-Leffler.

Figures 5.7.i and 5.7.ii on p. 106; from Henri Poincaré, *Sur le Problème des Trois Corps et les Équations de la Dynamique*, Acta Mathematica, **13** (1890); Courtesy of the Institut Mittag-Leffler.

Figure 5.8.i on p. 111; from Henri Poincaré, *Sur le Problème des Trois Corps et les Équations de la Dynamique*, Acta Mathematica, **13** (1890); Courtesy of the Institut Mittag-Leffler.

Figure 5.8.ii on p. 112; from Henri Poincaré, *Sur le Problème des Trois Corps et les Équations de la Dynamique*, Acta Mathematica, **13** (1890); Courtesy of the Institut Mittag-Leffler.

Figure 5.8.iii on p. 115; from Henri Poincaré, *Sur le Problème des Trois Corps et les Équations de la Dynamique*, Acta Mathematica, **13** (1890); Courtesy of the Institut Mittag-Leffler.

Figure 5.9.i on p. 123; from Henri Poincaré, *Sur le Problème des Trois Corps et les Équations de la Dynamique*, Acta Mathematica, **13** (1890); Courtesy of the Institut Mittag-Leffler.

Librairie Scientifique et Technique Albert Blanchard

Figure 2.3.iv on p. 25; from Henri Poincaré, *Les methodes nouvelles de la mechanique celeste* **I**, reprinted by Albert Blanchard, 1987, p. 109; Courtesy of Albert Blanchard, Librarie Scientifique et Technique.

The Johns Hopkins University Press

Figure 2.3.iii on p. 25; from *Collected mathematical works of G. W. Hill*, American Journal of Mathematics **1** (1878), p. 335; Courtesy of The Johns Hopkins University Press.

London Mathematical Society

> Figure 10.iii on cover and p. 226; from V. I. Arnold, *Small denominators and problems of stability of motion in classical and celestial mechanics*, Russian Mathematical Surveys, **18** (1963), p. 100; Courtesy of the London Mathematical Society.

The Royal Swedish Academy of Sciences

> Photograph of Hugo Gyldén on p. 140; from *Yearbook of the Royal Swedish Academy*, 1939; Courtesy of The Royal Swedish Academy of Sciences.

Springer-Verlag Berlin

> Figure 5.5.iii on p. 91; from June Barrow-Green, *Oscar II's prize competition and the error in Poincaré's memoir on the three body problem*, Archive for the History of Exact Sciences, 48, No. 2 (1994) p. 124; Courtesy of Springer-Verlag.

The Swedish National Art Museums

> Portrait of King Oscar II by Oskar Bjork on cover and p. 50; Courtesy of The Royal Palace Library.

The following figures and photographs are in the public domain.

> Figure 8.4.ii on p. 195; from Sir George Darwin, *Periodic Orbits*, Scientific Papers **IV**, p.12.

> Figure 9.2.i on p, 206; from Jacques Hadamard, *Œuvres de Jacques Hadamard*, vol. 2, 1968, pp. 741.

> Figure 9.2.ii on p, 206; from Jacques Hadamard, *Œuvres de Jacques Hadamard*, vol. 2, 1968, pp. 744.

> Figure 9.2.iii on p, 208; from Jacques Hadamard, *Œuvres de Jacques Hadamard*, vol. 2, 1968, pp. 758.

> Photograph of Alexander Liapunov on p. 178.

> Photograph of Karl Sundman on p. 188.

The American Mathematical Society holds copyright to the following.

> Figure 10.ii on p. 224; from Raoul Bott, *Marston Morse and his mathematical works*, Bull. Amer. Math. Soc. **3** (1980), p. 915.

> Photograph of George Birkhoff on p. 212.

Henri Poincaré

CHAPTER 1

Introduction

In November 1890 Henri Poincaré's memoir on the three body problem was published in the journal *Acta Mathematica* as the winning entry in the international competition honoring the 60th birthday of Oscar II, King of Sweden and Norway.[1]

Since its publication Poincaré's memoir has been lauded by succeeding generations of mathematicians as a milestone in the study and development of both celestial mechanics and dynamics. In 1902 F. R. Moulton wrote,

> *The methods employed by Poincaré were incomparably more profound and powerful than any previously used in Celestial Mechanics, and mark an epoch in the development of the science.*[2]

In 1925 George Birkhoff declared,

> *Le Problème de* [sic] *trois corps ... contained the first great attack upon the non-integrable problems of dynamics. ... Acta Mathematica has had many remarkable articles, but perhaps none of larger scientific importance than this one.*[3]

Rather more recently, Philip Holmes described the memoir as,

> *... the first textbook in the qualitative theory of dynamical systems ... .*[4]

Today the memoir is renowned both for providing the foundations for the author's celebrated three-volume *Les Méthodes Nouvelles de la Mécanique Céleste* and for containing the first mathematical description of chaotic behaviour in a dynamical system.

However, it has now come to light that this widely applauded paper is in fact very different from the version [P1] which actually won Oscar's prize. In his introduction to the published version [P2], Poincaré mentioned that he had revised the paper for publication, but gave no indication of the nature and extent of his alterations. Nevertheless, the discovery of a printed version of Poincaré's original memoir personally annotated by him has made it possible to reconstruct the exact nature of his alterations. Comparing [P1] and [P2], it can be seen that some of the principal results for which the paper is best known today are nowhere to be found in the original version, and, moreover, that the inclusion of these new results was not merely a consequence of Poincaré extending his existing material. Rather, their inclusion derives from Poincaré's discovery of a significant error, a discovery which he made only a few days before the memoir was due to be published in *Acta*. As a result of this discovery he was forced to rewrite a substantial part of the memoir,

---

[1] Poincaré [P2].
[2] Moulton [1914, *320*].
[3] Birkhoff [1925, *297*].
[4] Holmes [1990].

a process that significantly delayed the memoir's ultimate publication. Although
the existence of the error was known to some contemporaries (for example Moulton
mentions it in his obituary of Poincaré [4a] its seriousness has only been recognised
more recently. The significance of the error is evident from the fact that it was as
a result of its correction that Poincaré made his important discovery of *homoclinic
points*.

This book is an account of Poincaré's memoir, both from a mathematical and
an historical perspective. At the centre is a detailed study of the mathematics in
the memoir in that the two versions are compared and the error explained. The
memoir is put into an historical context through an examination of the mathematical environment in which it was created as well as of its place within Poincaré's
œuvre. From a study of its influence on later mathematical developments it is
seen that Poincaré's methodology marked an important change in approach to dynamical problems, a change which eventually resulted in the emergence of modern
dynamical systems theory. In addition, it is shown that several of the new and innovatory ideas that the memoir contained were later extended and developed both
in dynamics and other branches of mathematics.

With regard to the memoir's antecedents, two main strands are discussed. First,
we examine earlier work on the three body problem, in particular the contributions
of the 19th-century dynamical astronomers, especially those of G. W. Hill, whose
groundbreaking investigation into the theory of periodic solutions had a fundamental influence on Poincaré's research in this field. Second, we discuss Poincaré's own
researches relating directly to the memoir, the most important of these being his
pioneering study on the qualitative theory of differential equations, in which he laid
the foundations for many of the important ideas developed in the memoir.

The circumstances which led up to and surrounded the final publication of the
memoir are also described. The organisation of the Oscar competition, the original plans for which were first aired in 1884, was the responsibility of the Swedish
mathematician and editor of *Acta*, Gösta Mittag-Leffler. Mittag-Leffler engaged the
support of two of Europe's leading analysts, Charles Hermite and Karl Weierstrass,
to make up a commission to propose the questions and to judge the entries. Correspondence preserved at the Institut Mittag-Leffler shows that, despite appearances
to the contrary, the competition was in fact beleaguered by difficulties throughout,
difficulties that reached a climax with the discovery of Poincaré's error. An examination of the surviving documents provides an intriguing glimpse of life in the late
19th-century European mathematical community.

In describing and analyzing the mathematics in the memoir, we follow Poincaré's structure closely in order to describe clearly the significant differences between the two versions. Although Poincaré discussed several concepts in general,
in the main he concentrated on Hamiltonian systems with two degrees of freedom,
and in particular the restricted three body problem. In both versions he made a
clear distinction between theory and application.

The major part of the theory is taken up with the development of the topics
of invariant integrals and periodic solutions. Poincaré had touched on the subject
of invariant integrals in earlier papers, but it is here that he gives the first detailed
account of the theory and includes for the first time his famous recurrence theorem.
With regard to the periodic solutions, Poincaré had earlier indicated the potential

---

[4a] See Diacu and Holmes [1996, 48].

of these solutions for resolving qualitative questions in the theory of differential equations, but it was in this memoir that he successfully demonstrated what a powerful tool they could be. Of particular importance in this connection is his theory of asymptotic solutions.

The second part of the memoir is largely concerned with the application of the theory of asymptotic solutions to the restricted three body problem. It also contains Poincaré's theorem concerning the nonexistence of any new transcendental integrals of the restricted three body problem, together with his proof of the divergence of Lindstedt's series.

In his application of the theory of asymptotic solutions Poincaré proceeded by a series of approximations. This involved taking account of an increasing number of terms in the power series expansions for the solutions and considering the geometrical implications of the improved approximations. What eventually becomes clear is that, while it is the geometrical description of the asymptotic solutions that is most dramatically affected by the discovery and correction of the error, the essential changes that the correction entails are in fact due to a conjunction of errors arising in two separate parts of the preceding theoretical analysis.

Briefly, in his original account Poincaré did not draw a distinction between autonomous and nonautonomous Hamiltonian systems of differential equations. As a result he drew mistaken conclusions about the convergence of the series used to describe the asymptotic solutions of the problem. Originally he had believed that the series were convergent and led to asymptotic trajectories with behaviour which he could easily understand. In the revised memoir he showed that the series were actually asymptotic expansions, and, with his discovery of homoclinic points, he found that the behaviour of the trajectories was anything but easy to describe. In fact the behaviour was what today would be called chaotic. Thus, contrary to what is sometimes thought, Poincaré did not win the Oscar prize for his discovery and analysis of the behaviour of what he called *doubly asymptotic* solutions (and later called homoclinic solutions), that is, the 'chaotic' trajectories, but rather for the underlying theory which eventually led to his correct description of these solutions.

Given the originality of Poincaré's methodology, it is not surprising to find that the memoir did not meet with universal approval from his contemporaries. Although the overall response was one of approbation and admiration, the novelty of his ideas, combined with his capacity for conciseness, also evoked feelings of confusion and bewilderment. Hugo Gyldén, a Finnish dynamical astronomer, was the loudest of the dissenters, but confusion was also rife among mathematicians. Even such luminaries as Weierstrass and Hermite struggled in their efforts to understand Poincaré's mathematics.

After the publication of the memoir, Poincaré continued working on problems in celestial mechanics. A stream of short papers followed, ranging from specific corrections and criticisms of the work of other mathematicians and astronomers to articles of a general nature on the stability of the solar system. This period of prodigious productivity was crowned by his *Méthodes Nouvelles*, which was published between 1892 and 1899, the first and last volumes being largely an elaboration and refinement of the memoir.

In two later papers Poincaré developed his ideas about the existence of periodic solutions in the three body problem within a topological framework. The first of these, which appeared in 1905 and was his first on periodic solutions after an interval of more than ten years, was a study of geodesics on a convex surface. This appeared shortly after the final paper in his fundamental series on the study of topology (or *analysis situs*, as it was then called). The second paper, which was published in 1912 shortly before he died, contained his famous 'last geometric theorem', the complete proof of which had eluded him. The proof was supplied shortly afterwards by the young American mathematician George Birkhoff.

The latter part of this book is concerned with the influence of the memoir on the progress of the three body problem, as well as its role in the foundation of dynamical systems theory. In general, the focus is on work done in the period up to 1920. The date 1920 was chosen because, by this time, not only had a function theoretical proof to the three body problem been obtained, but also dynamical systems theory had started to move into a new phase no longer centred around problems in celestial mechanics. This move can largely be attributed to the inadequacy of computing techniques, which meant that verification of the accuracy of predictions generated by the theory was not feasible. Furthermore, not only did this lack of computing power help to engender a move towards more generalised problems, but also astronomy itself was changing. The announcement of Einstein's general theory of relativity in 1915 and the rise of quantum mechanics in the 1920s created additional diversions away from some of the traditional problems in celestial mechanics, with mathematicians eager to find applications for the new methods.

As a preliminary to the discussion of [P2]'s influence, we extend the context in which Poincaré's work was produced by considering Alexander Liapunov's qualitative study of stability theory. Although Liapunov's famous memoir, first published in 1890, was produced independently of [P2], it does provide an interesting alternative account of one of the topics discussed by Poincaré.

As far as the three body problem itself is concerned, the decades following 1889 were dominated by the search for ways to regularise the equations. It will be seen that the key figures in this part of the story were Paul Painlevé, who, in his Stockholm lectures published in 1895, made the first major advance; Tullio Levi-Civita, who worked extensively on the problem for many years; and Karl Sundman, who was responsible for its resolution in 1912. Also important in this era was the quantitative work of Sir George Darwin. Darwin worked unstintingly on this aspect of the problem, engaging in a detailed study of periodic orbits.

With regard to the more general problems of dynamics, Poincaré's methods, being characterised by a global geometric viewpoint, led to the opening up of a new qualitative approach to the subject. This is traced in a discussion of the work of both Jacques Hadamard and George Birkhoff, each of whom were greatly influenced by Poincaré and professed the greatest admiration for his work.

Hadamard's ideas were presented in two fundamental papers concerning geodesics on surfaces which were published in 1897 and 1898. The first, for which he was awarded the *Prix Bordin de l'Académie des Sciences*, deals with the case when the surfaces are everywhere of positive curvature, while the second considers surfaces of negative curvature. Of special note in these two papers is Hadamard's emphasis on the importance of topology for the qualitative study of differential equations, which reflects a very visible response to Poincaré.

George Birkhoff made outstanding progress in the field of dynamics and the general theory of orbits. He was especially influenced by Poincaré's *Méthodes Nouvelles*, and recorded particular successes in geometrical aspects of dynamics, notably in providing a proof for Poincaré's last geometric theorem. Birkhoff, in his own work on the restricted three body problem and dynamical systems, expanded Poincaré's ideas to become one of the founders of modern dynamics.

Since Poincaré's role in the foundation of modern chaos theory has been amply dealt with elsewhere, it is not discussed here, but a forward look is provided by a glance at the work of some later mathematicians whose research into particular aspects of dynamical systems derives ultimately from Poincaré. These mathematicians include Marston Morse, a student of Birkhoff's, who took up both Birkhoff's ideas and those of Hadamard and who thought of himself as a mathematical descendant of Poincaré. Morse understood topology through dynamics and in 1917 wrote papers on dynamics and geodesic flow that were the starting point of his work on symbolic dynamics. Poincaré's ideas also proved seminal to the Russian mathematician Melnikov, who in 1963 developed a method for the detection of chaos and nonintegrability in perturbed Hamiltonian systems. Finally, in the 1950s and 1960s, Kolmogorov, Arnold, and Moser tackled specific questions raised by Poincaré's results in the field of periodic solutions. Their brilliant investigations into the existence of quasi-periodic solutions of Hamiltonian systems led to the development of their now renowned KAM theory.

Throughout this book, references will be cited in the [year] or [year, *page number*] form. All page numbers in references to works of Poincaré, Birkhoff, Darwin, Hadamard, and Hill will refer to the collected works where applicable. The references, together with a name and date index, are collected at the end of the book. Unless otherwise stated, all translations are my own.

CHAPTER 2

Historical Background

2.1. Introduction

The three body problem, which was described by Whittaker as *"the most celebrated of all dynamical problems"*[5] and which fulfilled for Hilbert the necessary criteria for a good mathematical problem,[6] can be simply stated: three particles move in space under their mutual gravitational attraction; given their initial conditions, determine their subsequent motion. Like many mathematical problems, the simplicity of its statement belies the complexity of its solution. For although the one and two body problems can be solved in closed form by means of elementary functions, the three body problem is a complicated nonlinear problem, and no similar type of solution exists.

Apart from its intrinsic appeal as a simple-to-state problem, the three body problem has a further attribute which has contributed to its attraction for potential solvers: its intimate link with the fundamental question of the stability of the solar system. Over the years attempts to find a solution spawned a wealth of research, and between 1750 and the beginning of this century more than 800 papers relating to the problem were published, invoking a roll call of many distinguished mathematicians and astronomers.[7] And hence, as is often the case with such problems, its importance is now perceived as much in the mathematical advances generated by attempts at its solution as in the actual problem itself. These advances have come in many different fields, including, in recent times, the theory of dynamical systems.

At the beginning of this century a Finnish mathematical astronomer, Karl Sundman, mathematically "solved" the problem by providing a convergent power series solution valid for all values of time. However, since Sundman's solution gives no qualitative information about the behaviour of the system, and the rate of convergence of the series is considered to be too slow to be of any real practical use, it leaves plenty of issues surrounding the problem unresolved.[8]

---

[5]Whittaker [1937,*339*].

[6]See the introduction to Hilbert's famous speech on mathematical problems given at the Paris Congress in 1900, published in *Gottinger Nachrichten* **1900**, *253-297*, *Archiv der Mathematik und Physik* (3) **1** (1901), *44-63* and *213-237*, and translated by Mary Winston Newson for the *Bulletin of the American Mathematical Society* **8** (1902), *437-479*.

[7]Whittaker [1937, *339*].

The history of the three body problem has been well documented, most notably by Gautier [1817], Cayley [1862], Whittaker [1899], Lovett [1912], and Marcolongo [1919]. Whittaker's famous report describes the situation from 1868 to 1898, while that of Lovett is concerned with the following decade. Marcolongo's book is a full account, superbly referenced, and deserves to be better known.

[8]See Chapter 8 for a discussion of Sundman's solution.

To clarify the mathematical difficulties associated with the problem and to provide a context for historical development, we will begin with a modern mathematical description.[9]

2.2. Mathematical description of the three body problem

2.2.1. The differential equations of the problem.

Let P_i represent the three particles with masses m_i, distances $P_i P_j = r_{ij}$, and coordinates $q_{ij}(i, j = 1, 2, 3)$ in an inertial coordinate system. The equations of motion are

(1)
$$\frac{d^2 q_{1i}}{dt^2} = k^2 m_2 \frac{(q_{2i} - q_{1i})}{r_{12}^3} + k^2 m_3 \frac{(q_{3i} - q_{1i})}{r_{13}^3}$$

$$\frac{d^2 q_{2i}}{dt^2} = k^2 m_1 \frac{(q_{1i} - q_{2i})}{r_{12}^3} + k^2 m_3 \frac{(q_{3i} - q_{2i})}{r_{23}^3}$$

$$\frac{d^2 q_{3i}}{dt^2} = k^2 m_1 \frac{(q_{1i} - q_{3i})}{r_{13}^3} + k^2 m_2 \frac{(q_{2i} - q_{3i})}{r_{23}^3} \qquad (i = 1, 2, 3),$$

where k is the gravitational constant. The problem is therefore described by nine second-order differential equations.

If the units are chosen such that k^2 is equal to one, the force of attraction between the ith and jth particles becomes $m_i m_j / r_{ij}^2$, and the corresponding term in the potential energy becomes $-m_i m_j / r_{ij}$. The potential energy V of the whole system is then given by

$$V = -\frac{m_2 m_3}{r_{23}} - \frac{m_3 m_1}{r_{31}} - \frac{m_1 m_2}{r_{12}}.$$

Writing

$$p_{ij} = m_i \frac{dq_{ij}}{dt}$$

and

$$H = \sum_{i,j=1}^{3} \frac{p_{ij}^2}{2m_i} + V,$$

the equations take the Hamiltonian form

$$\frac{dq_{ij}}{dt} = \frac{\partial H}{\partial p_{ij}}, \qquad \frac{dp_{ij}}{dt} = -\frac{\partial H}{\partial q_{ij}},$$

which is a set of 18 first-order differential equations.

So for a closed solution to the problem the system needs 18 independent integrals. However, it is only possible to find 12 such integrals, and the system can therefore only be reduced to one of order six. As will be shown below, this is achieved through the use of the so-called ten classical integrals—the six integrals of the motion of the centre of mass, the three integrals of angular momentum, and the energy integral—together with the elimination of the time and the elimination of what is called the ascending node. Bruns [1887] proved that, when the rectangular coordinates are chosen as dependent variables, the ten classical integrals are the only independent algebraic integrals of the problem, and all others can be formed by a combination of them. It will be seen later how Poincaré extended Bruns'

---

[9]The general mathematics of the three body problem is discussed in many classic texts on both analytical dynamics and celestial mechanics. Good examples are to be found in Brouwer and Clemence [1961], Moulton [1914], Pars [1968], Whittaker [1937], and Wintner [1941]. For a detailed treatment see Marchal [1990].

result by establishing that, provided the masses of two of the bodies are very small when compared with the third, no new single-valued transcendental integrals of the problem exist.

2.2.2. Reduction to 6th order. If the ijth equation of equations (1) is multiplied by m_i, a summation can be performed to give three equations

$$\sum_{i=1}^{3} m_i \frac{d^2 q_{ij}}{dt^2} = 0, \qquad (j = 1, 2, 3).$$

These equations can be integrated twice to give the equations

$$\sum_{i=1}^{3} m_i q_{ij} = A_j t + B_j, \qquad (j = 1, 2, 3),$$

in which the A_j and B_j are constants of integration. These equations show that the centre of mass of the three particles either remains at rest or moves uniformly in space in a straight line. This is as expected since there are no forces acting except the mutual attractions of the particles. The six constants serve to describe the motion of the centre of mass in the original arbitrary inertial coordinate system and play no part in the motion of the bodies about the centre of mass.

If in the first set of equations (1) the first equation is multiplied by $-q_{12}$, the second equation by $-q_{22}$ and the third equation by $-q_{32}$, and in the second set the first equation is multiplied by q_{11}, the second equation by q_{21}, and the third equation by q_{31}, and these two sets are added together, then this gives

$$\sum_{i=1}^{3} m_i q_{i1} \frac{d^2 q_{i2}}{dt^2} - \sum_{i=1}^{3} m_i q_{i2} \frac{d^2 q_{i1}}{dt^2} = 0,$$

and two similar equations can be obtained by a cyclic change of the variables. The three equations can then be integrated to give

$$\sum_{i=1}^{3} m_i \left(q_{i2} \frac{dq_{i3}}{dt} - q_{i3} \frac{dq_{i2}}{dt} \right) = C_1$$

$$\sum_{i=1}^{3} m_i \left(q_{i3} \frac{dq_{i1}}{dt} - q_{i1} \frac{dq_{i3}}{dt} \right) = C_2$$

$$\sum_{i=1}^{3} m_i \left(q_{i1} \frac{dq_{i2}}{dt} - q_{i2} \frac{dq_{i1}}{dt} \right) = C_3.$$

These equations represent the conservation of angular momentum for the system. That is, they show that the angular momentum of the three particles around each of the coordinate axes is constant throughout the motion.

Looked at geometrically, the terms in brackets represent the projections of the areal velocities of the various bodies upon the three coordinate planes, and hence the integrals are also known as the integrals of area.[10]

Since

$$\frac{\partial}{\partial q_{ij}} \left(\frac{1}{r_{ik}} \right) = -\frac{q_{kj} - q_{ij}}{r_{ik}^3},$$

---

[10]Relative to a fixed origin, the areal velocity of a point is the area swept out by the radius vector per unit time.

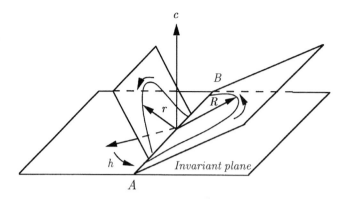

c: *total angular momentum vector*
AB: *lines of nodes*
r: *radius vector of inner orbit*
R: *radius vector of exterior orbit*
h: *longitude of the node of the inner orbit*

FIGURE 2.2.i

the equations of motion can be written in the form

$$m_i \frac{d^2 q_{ij}}{dt^2} = -\frac{\partial V}{\partial q_{ij}}.$$

Multiplying by $\frac{dq_{ij}}{dt}$ and summing gives, since V is a function of the coordinates only,

$$\sum_{i,j=1}^{3} p_{ij} \frac{d^2 q_{ij}}{dt^2} = -\frac{dV}{dt}.$$

This equation can then be integrated to give

$$\sum_{i,j=1}^{3} \frac{p_{ij}^2}{2m_i} = -V + C,$$

where C is a constant of integration. Furthermore, since the left-hand side of the equation represents the kinetic energy T of the system, the integral can be put into the form $T - F = C$, which expresses the conservation of energy.[11]

Two final reductions can then be made to the order of the system. First, the time can be eliminated by using one of the dependent variables as an independent variable, and, second, a reduction can be made by the so-called elimination of the nodes, a procedure first made explicit by Jacobi [1843], although indicated in Lagrange [1772].

The first stage of Jacobi's process is to make a linear change of variables which effectively changes the configuration to one in which two "fictitious" bodies orbit a third. Since the change of variable is linear, the form of the integrals of angular momentum is unchanged and the total angular momentum vector **c** remains constant and perpendicular to an invariant plane (see *Figure 2.2.i*). Jacobi showed that the

---

[11]Sometimes called the "Vis Viva" integral.

intersection between the orbital planes of the two bodies remains parallel to this invariant plane. Hence the difference in longitude between the ascending nodes is always π radians.

Thus through use of the classical integrals and these last two integrals, the original system of order 18 can be reduced to a system of order six. Furthermore, this result can be generalised to the n body problem. In this case the differential equations constitute a system of order $6n$. By using the same integrals this system can be reduced to a system of order $(6n - 12)$.

2.2.3. The restricted three body problem. A special case of the three body problem which has featured prominently in research due to its simplified form and its practical applications is what is known today as the "restricted" three body problem.[12] In this formulation two of the bodies revolve around their centre of mass in circular orbits under the influence of their mutual gravitational attraction, and hence form a two body system in which their motion is known. A third body (generally known as the planetoid), assumed massless with respect to the other two, moves in the plane defined by the two revolving bodies and, while being gravitationally influenced by them, exerts no influence of its own. The problem is then to ascertain the motion of the third body.

This particular case of the three body problem is the simplest one of importance and, in the context of Poincaré's memoir, is especially significant since most of his results pertain to this formulation. Apart from its simplifying characteristics, it also provides a good approximation for real physical situations, as, for example, in the problem of determining the motion of the moon around the earth, given the presence of the sun. In this instance, the problem is almost circular (the eccentricity of the earth's orbit is approximately 0.017) and, almost planar (both the earth's orbit and the moon's orbit are nearly in the plane of the ecliptic), and the values of the mass ratios and the mean distances between the bodies satisfy the conditions.

The differential equations of motion for the restricted problem can be derived as follows. Let $m_1 = 1 - \mu$ and $m_2 = \mu$ denote the masses of the two finite bodies, choose the unit of distance so that the constant distance between the two finite bodies a is unity, and choose the unit of time so that the gravitational constant k^2 is also unity. If the coordinates of m_1, m_2, and the planetoid P are (ξ_1, η_1), (ξ_2, η_2), and (ξ, η) respectively and

$$r_1 = \sqrt{(\xi - \xi_1)^2 + (\eta - \eta_1)^2}$$
$$r_2 = \sqrt{(\xi - \xi_2)^2 + (\eta - \eta_2)^2},$$

then the equations of motion of the planetoid are

(2)
$$\frac{d^2\xi}{dt^2} = -(1 - \mu)\frac{(\xi - \xi_1)}{r_1^3} - \mu\frac{\xi - \xi_2}{r_2^3}$$
$$\frac{d^2\eta}{dt^2} = -(1 - \mu)\frac{(\eta - \eta_1)}{r_1^3} - \mu\frac{\eta - \eta_2}{r_2^3}.$$

By Kepler's third law, the mean angular motion n of the finite bodies is

$$n = k\frac{\sqrt{(1 - \mu) + \mu}}{a^{3/2}},$$

which, due to the way the units have been chosen, is equal to unity.

---

[12]The first use of the term *problème restreint* occurs in Poincaré [MN III, *69*].

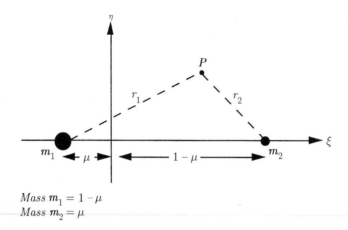

Mass $m_1 = 1 - \mu$
Mass $m_2 = \mu$

FIGURE 2.2.ii

If the motion of the bodies is now referred to a new system of axes x and y having the same origin as the old axes but rotating in the $\xi\eta$ plane in the direction in which the finite bodies move with uniform angular velocity, then the coordinates of the planetoid in the new system are defined by

$$(3) \qquad\qquad \xi = x \cos t - y \sin t$$
$$\eta = x \sin t + y \cos t,$$

with similar sets of equations for (ξ_1, η_1) and (ξ_2, η_2) (*Figure 2.2.ii*).

Differentiating equations (3) twice, substituting the resulting expressions in equations (2), multiplying the two equations by $\cos t$ and $\sin t$ respectively and adding, and then multiplying them by $-\sin t$ and $\cos t$ respectively and adding, gives

$$\frac{d^2 x}{dt^2} - 2\frac{dy}{dt} = x - (1 - \mu)\frac{(x - x_1)}{r_1^2} - \mu\frac{(x - x_2)}{r_2^3}$$

$$\frac{d^2 y}{dt^2} + 2\frac{dx}{dt} = y - (1 - \mu)\frac{(y - y_1)}{r_1^2} - \mu\frac{(y - y_2)}{r_2^3}.$$

Since it is always possible to choose the direction of the x axis so that the two finite bodies lie on it, in which case $y_1 = y_2 = 0$, the equations then become

$$(4) \qquad\qquad \frac{d^2 x}{dt^2} - 2\frac{dy}{dt} = x - (1 - \mu)\frac{(x - x_1)}{r_1^2} - \mu\frac{(x - x_2)}{r_2^3}$$

$$\frac{d^2 y}{dt^2} + 2\frac{dx}{dt} = x - (1 - \mu)\frac{y}{r_1^2} - \mu\frac{y}{r_2^3}.$$

These differential equations, which have the important property that they do not involve t explicitly, are the equations of motion of the planetoid with respect to the rotating coordinates. Since they are a set of two second-order equations they represent a system of order four. However, as Jacobi showed, they do admit a solution, now known as the Jacobian integral, which reduces the system to one of order three.[13]

---

[13] Jacobi [1836] first announced his integral in an inertial coordinate system.

To find the Jacobian integral first define a function U by

$$U = \frac{1}{2}(x^2 + y^2) + \frac{1-\mu}{r_1} + \frac{\mu}{r_2}.$$

Equations (4) can then be written

(5)
$$\frac{d^2x}{dt^2} - 2\frac{dy}{dt} = \frac{\partial U}{\partial x}$$

$$\frac{d^2y}{dt^2} + 2\frac{dx}{dt} = \frac{\partial U}{\partial y}.$$

Multiplying these by $2\dfrac{dx}{dt}$ and $2\dfrac{dy}{dt}$ respectively and adding gives

(6)
$$2\frac{dx}{dt}\frac{d^2x}{dt^2} + 2\frac{dy}{dt}\frac{d^2y}{dt^2} = 2\frac{dx}{dt}\frac{\partial U}{\partial x} + 2\frac{dy}{dt}\frac{\partial U}{\partial y},$$

which is an exact differential since U is a function of x and y alone.

Integrating equation (6) gives

$$\left(\frac{dx}{dt}\right)^2 + \left(\frac{dy}{dt}\right)^2 = 2U - C,$$

where C is a constant of integration and the left-hand side is the square of the velocity of the planetoid in the rotating frame. If the latter is denoted V^2 then

$$V^2 = 2U - C,$$

and this solution is the Jacobian integral of the restricted three body problem. It is sometimes misleadingly called the energy or relative energy integral, but this terminology is erroneous since the integral does not express conservation of energy. The total energy of the original two body system remains constant but that of the planetoid does not, and so the total energy is not constant. As Szebehely points out, the solution should be regarded purely as an integral of the equations of motion of the restricted three body problem using rotating coordinates.[14]

The equations (2) can also be put into Hamiltonian form, and since Poincaré used this form of the equations in his memoir, it is useful to see how it is derived.

If the equations are first put in the form

$$\frac{d^2\xi}{dt^2} = \frac{\partial F}{\partial \xi}, \qquad \frac{d^2\eta}{dt^2} = \frac{\partial F}{\partial \eta},$$

where $F = \dfrac{m_1}{r_1} + \dfrac{m_2}{r_2}$, then if $\eta = q_1$, $\dfrac{d\eta}{dt} = p_1$, and $\dfrac{d\eta}{dt} = p_2$, the Hamiltonian form of the equations is

$$\frac{dq_i}{dt} = \frac{\partial H}{\partial p_i}, \qquad \frac{dp_i}{dt} = -\frac{\partial H}{\partial q_i} \qquad (i = 1, 2),$$

where $H = \frac{1}{2}(p_1^2 + p_2^2) - F$.

The function F is not only a function of the variables q_i but is also a function of the time t, and hence

$$H = constant$$

is not a solution of the system. But, if, as in the previous formulation, a transformation is made such that the axis of rotation and a line perpendicular to this through the centre of gravity of the two bodies become the coordinate axes, then

---

[14]See Szebehely [1967, *12-13, 38*].

a solution can be found. As Whittaker has shown [1937, *354*], this can be done by applying the contact transformation defined by the equations

$$q_i = \frac{\partial W}{\partial p_i}, \qquad P_i = \frac{\partial W}{\partial Q_i},$$

where

$$W = p_1(Q_1 \cos nt - Q_2 \sin nt) + p_2(Q_1 \sin nt + Q_2 \cos nt),$$

and n is the uniform angular velocity of the rotating axis which, due to the choice of units, is equal to unity.[15]

The new Hamiltonian, which has no explicit dependence on time, is then given by

$$H' = H - \frac{\partial W}{\partial t},$$

and

$$H' = constant$$

is a solution of the system corresponding to the Jacobian integral.

2.3. History of the three body problem

2.3.1. Origin of the three body problem. Since bodies in the solar system are approximately spherical and their dimensions extremely small when compared with the distances between them, they can be considered as point masses. Hence the origin of the problem can be thought of as being synonymous with the foundation of modern dynamical astronomy. This part of celestial mechanics, which connects the mechanical and physical causes with the observed phenomena, began with the introduction of Newton's theory of gravitation. From the time of the publication of the *Principia* in 1687, it became important to verify whether Newton's law alone was capable of rendering a complete understanding of how celestial bodies move in space. In order to pursue this line of investigation, it was necessary to ascertain the relative motion of n bodies attracting one another according to the Newtonian law.

Newton himself had geometrically solved the problem of two bodies for two spheres moving under their mutual gravitational attraction [1934, **I**, *section XI*], and in 1710 Johann Bernoulli had proved that the motion of one particle with respect to the other is described by a conic section.[16] In 1734 Daniel Bernoulli won a French Academy prize for his analytical treatment of the two body problem,[17] and the problem was solved in detail by Euler [1744]. Meanwhile work was already in progress on the higher dimensional problem. Driven by the needs of navigation for knowledge about the motion of the moon, researchers scrutinized the system formed by the sun, the earth and the moon, and the lunar theory (as the study of the moon's motion came to be called) quickly dominated the early research into the problem.

---

[15]The idea of contact transformations originated with Sophus Lie in the early 1870s. See Hawkins [1992].

[16]See Wintner [1941, *420*].

[17]See Kline [1972, *492*].

2.3.2. Early attempts at solution. The investigations arising from a search for a solution to the problem led in two directions: those which were concerned with finding general theorems concerning the motion, and those which were searching for good approximations for solutions that would hold for a given period of time starting from an instant at which data was available.

Newton was the first to treat the problem, and he achieved results in both types of investigation. On the one hand, having previously shown that the centre of mass of n bodies moves with uniform speed in a straight line, he made a general investigation into the motion of attracting bodies,[18] while on the other, using an essentially geometric approach to the method of variation of parameters, he applied perturbation theory to the motion of the moon. Having treated the motion of the moon about the earth and obtained an elliptical orbit, he considered the effect of the sun on the moon's orbit by taking account of variations in the latter. However, the calculations caused him great difficulties, and his computation for the motion of the lunar apsides[19] gave a value which was approximately half that of the observed value—a fact which he encapsulated in later editions of the *Principia* in the brief sentence: *"The apse of the moon is about twice as swift."*[20] Indeed, the problems he encountered were such that he was prompted to remark to the astronomer John Machin that *"... his head never ached but with his studies on the moon."*[21]

During the 18th century, the gradual recognition of the power of analytic methods meant that dynamics in general, and celestial mechanics in particular, began to break free from the constraints of geometry. With this freedom came the realisation of the impossibility of finding a closed solution in terms of elementary functions. Clairaut in [1747] announced the first successful approximate resolution to the lunar problem by using infinite series solutions to an improved simplification of the differential equations. However, the difficulty he had in explaining the motion of the lunar perigee was such that he even considered a modification to the inverse square law. But in [1749], by carrying his original approximation one step further, he reached results which almost accounted for the motion, and in 1752 his *Théorie de la Lune* won the St. Petersburg Academy prize. The value of Clairaut's methods was amply confirmed by his prediction of the date of the perihelion passage of Halley's comet in 1759, which was accurate to within one month, almost exactly the margin of error he had allowed himself. (In 1872 it was discovered that Newton, in an unpublished manuscript in the Portsmouth Collection, had in fact corrected his original calculations for the motion of the lunar perigee by including the second-order perturbations, although the correction was completely unknown until the manuscript's discovery.)[22]

Meanwhile, Euler [1748] was the first to use the method of variation of parameters to treat perturbations of planetary motion, and in [1753] he published his first lunar theory. His second lunar theory [1772], which, jointly with a memoir by Lagrange, shared the *Prix de l'Académie de Paris*, contained many new and

---

[18]Newton [1934, **I**, Section XI]

[19]The lunar apsides, also called the apogee and the perigee, are the two points in the orbit of the moon at which it is at its greatest and least distance from the earth respectively.

[20]Newton [1934 **I**, *147*]

[21]Keynes MSS 130.6, Book 3; 130.5, Sheet 3 - Newton Ms, in the Keynes collection in the library of King's College, Cambridge. See Westall [1980, *544*].

[22]Catalogue of the Portsmouth Collection of Books and Papers written by or belonging to Sir Isaac Newton, Cambridge, 1888, xi-xiii, xxvi-xxx, section 1, div ix, numbers 7, 12. See Newton [1934, *649*].

important features, including the first formulation of the restricted problem based on a rotating coordinate system together with particular solutions.[23]

Lagrange's prize-winning memoir [1772] was an analysis of the three body problem in which he showed that the problem could be reduced from a system of order 18 to a system of order seven.[24] His method was first to determine the mutual distances between the bodies, then to determine the plane of the triangle in space, and finally to determine the orientation of the triangle in the plane. In addition, he also found two types of particular solutions to the general problem. Jacobi in [1843], unaware of Lagrange's work, achieved an explicit reduction of the general problem to a sixth-order system through the elimination of the nodes.

As far as the restricted problem was concerned, although Poincaré was the first to coin the phrase, Euler [1772] was the first to propose the problem's formulation. A significant insight into the formulation was provided by Jacobi [1836], who showed that it could be represented by a system of fourth-order differential equations, one solution of which (the Jacobian integral) could be found. G. W. Hill, the American mathematician and astronomer, was the first to show that an important application of the Jacobian integral is its use in the establishment of the regions of motion for the planetoid [1878]. Hill's idea was later used to great effect by George Darwin [1897] in his quantitative investigations into periodic orbits (see Chapter 8).

2.3.3. Particular solutions.

Particular solutions are those solutions in which the geometric configuration of the three bodies remains invariant with respect to time. Thus, either the configuration simply rotates in its own plane around the centre of mass, or an expansion or contraction takes place in which the mutual distances between the three bodies remain in fixed ratios to each other. If there is no change in scale, the solutions are called stationary.

The particular solutions found by Euler in his revolving coordinate system were collinear solutions, while those of Lagrange took the form of both collinear and equilateral configurations. In the collinear case, the bodies are all set in motion from positions on a straight line, and, given appropriate initial conditions, they continue to stay on that line while the line rotates in a plane about the centre of mass of the bodies. In the equilateral case, the initial positions of the three bodies are at the vertices of an equilateral triangle, and the bodies continue to move as though attached to the triangle that rotates about the centre of mass.

Associated with these particular solutions are five equilibrium points, also called Lagrangian or libration points, L_1 to L_5 (*Figure 2.3.i*).

From a physical point of view, the Lagrangian points are the points where the forces acting on the third body in a rotating system are balanced, so there is no motion relative to the rotating system, and only the gravitational and centrifugal forces have to be considered. Lagrange proved the existence of triangular equilibrium points in the Sun-Jupiter system, and in so doing predicted the presence of the Trojan asteroids (*Figure 2.3.ii*), observational verification of which was not made until 1906 when Max Wolf discovered Achilles.

---

[23]Whittaker [1899, *123*] incorrectly credits the first discussion of the restricted problem to Jacobi.

[24]According to Whittaker [1937, *341*], the reduction from seventh to sixth order through the elimination of the nodes is implicit in Lagrange's memoir.

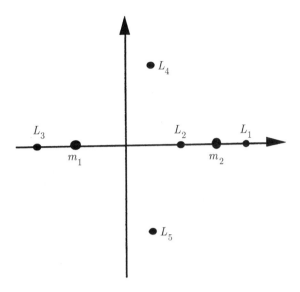

FIGURE 2.3.i

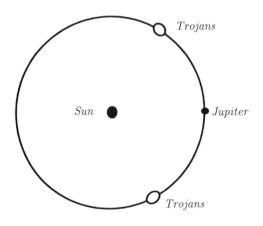

FIGURE 2.3.ii

2.3.4. "Small divisors". In planetary perturbation theory the disturbing function can be expanded in a series of periodic terms which when integrated produces terms of the form

$$\frac{A_{ij}}{(jn_1 + kn_2)} \sin[(jn_1 + kn_2)t + B_{jk}],$$

where j and k are integers, n_1 and n_2 are the mean motions of the planets, A and B are constants, and the order of magnitude of A diminishes rapidly as j and k increase. And these terms arise in the expressions for the Keplerian elements. If the mean motions are commensurable then terms with argument $[(jn_1 + kn_2)t + b]$ contribute to the secular term, but if they are incommensurable then the denominator can become arbitrarily close to zero. In practice the mean motions are determined by observation and given to a certain number of significant figures; hence it is always possible to find values for j and k for which the denominator is arbitrarily small. Nevertheless, incommensurability only really gives rise to problems when the mean motions can be closely approximated by the ratio of two small integers, since in this case the amplitude A remains large. It is this situation that leads to the phenomena of *small divisors*, a notorious problem in celestial mechanics. In more general dynamical terms the problem of small divisors translates into the problem of resonance caused by near commensurability of the natural frequencies.

In 1785 Laplace announced the resolution of several of the outstanding anomalies in the theory of the solar system, one of which concerned the observed deviations from Keplerian orbits of the planets Jupiter and Saturn. He had discovered a long-period inequality in the motions of these two planets that was due to terms of the third order in the eccentricities. This long-period inequality of Jupiter and Saturn is a particular example of the problem of small divisors. In this case the mean motions are approximately in the ratio of 5:2, and the expansion of the disturbing function gives a term with $j = -2$ and $k = 5$ in its argument, which is of the order of three in the eccentricities. Although the value of the term in the disturbing function is very small, its effect is striking in the perturbation of the mean longitude since this contains the square of its reciprocal. Furthermore, the period of the perturbation is $2\pi/(jn_1 + kn_2)$, which in this case works out to be approximately 900 years, hence the term long-period inequality.

2.3.5. Series representation. By the middle of the 19th century, it was clear that the possibility of finding a closed solution to the problem was becoming increasingly unlikely. Consequently, the objective became to improve the approximations that resulted from the solution of the differential equations being given as infinite series. This involved attempting to eliminate the *secular* terms, that is, those terms which increase or decrease indefinitely with time (the existence of which ultimately leads to an entirely new configuration of the system), in order to try to confine the expansion to series in which the time only occurs within the arguments of the periodic terms.[25]

However, in conjunction with the progress being made, a new complication evolved. This complication derived from the fact that those working on the problem were primarily astronomers rather than mathematicians and, as such, were interested in numerical rather than theoretical research. Critical to the problem was that this difference of emphasis, numerical versus theoretical, in respect of the two disciplines had resulted in a different understanding of the meaning of convergence. For the astronomers a series was considered to be convergent if the terms they had calculated decreased rapidly, regardless of the fact that they had no knowledge of the behaviour of the subsequent terms. For the mathematicians a series was considered to be convergent only if it was rigorously proved to be so.[26] The problem was that this difference was not fully recognised, and only later did it become clear that most of the series being proposed as solutions were not convergent in a rigorous mathematical sense.

---

[25] It is easy to see that the presence of secular terms can be extremely misleading: for example, if the true solution contains a term involving $\sin at$, then the series will contain terms in odd powers of at.

[26] Poincaré gives a nice illustration of this difference in [MNII]: see section 7.2.2 below.

For all practical purposes the calculation of the first terms provided a very satisfactory approximation, but for those occasions when the series were intended for the establishment of rigorous theoretical results, such as the question of the stability of the solar system, the divergence of the series posed a serious problem. Furthermore, if the series were divergent, then, in general, they would not be capable of providing an arbitrarily close approximation. Poincaré, while giving due credit to the value of the results of the astronomers—he singled out the achievements of Delaunay, Lindstedt and Gyldén as worthy of special mention—was the first to prove the general divergence of their series.

2.3.6. Delaunay. Delaunay [1860, 1867], by using the method of variation of parameters to form a purely trigonometric series that formally satisfied the equations of motion, was the first to complete a total elimination of the secular terms in the problem of lunar theory. Beginning with the three-dimensional elliptical restricted three body problem,[27] he completely developed the disturbing function R up to the seventh order of the small parameters and then repeatedly applied canonical transformations in order to eliminate the more important terms of R. The number of calculations involved in the project was immense, and it took Delaunay over twenty years to complete. He announced an outline of the principle of his method in [1846], but his final results, which occupied two large volumes, did not appear until 1860 and 1867.

Delaunay's variables were three canonically related pairs of orbital elements which gave the equations of motion Hamiltonian form. If a is the semi-major axis, e the eccentricity, i the inclination of the orbit to a fixed plane, μ the sum of the masses of the bodies whose relative motion is being considered, l the mean anomaly,[28] g the angular distance of the lower apsis from the ascending node, and h the longitude of the ascending node, then putting $L = \sqrt{\mu a}$, $G = L\sqrt{(1 - e^2)}$, $H = G \cos i$, and $F = R + \dfrac{\mu^2}{2L^2}$, the equations are[29]

$$\frac{dL}{dt} = \frac{\partial F}{\partial l}, \quad \frac{dG}{dt} = \frac{\partial F}{\partial g}, \quad \frac{dH}{dt} = \frac{\partial F}{\partial h},$$

$$\frac{dl}{dt} = -\frac{\partial F}{\partial L}, \quad \frac{dg}{dt} = \frac{\partial F}{\partial G}, \quad \frac{dh}{dt} = -\frac{\partial F}{\partial H}.$$

The Hamiltonian F was expanded as an infinite series (Delaunay's expansion contained 320 terms), and, using successive canonical transformations, the important periodic terms of the disturbing function R were eliminated term by term and the equations solved.

Although the method was accurate up to one second of arc, its practical use was hampered by the progressively increasing complexity of the expressions involved, combined with the slow convergence of its series. Nevertheless, since its publication Delaunay's formulation in terms of canonical systems of elements has been recognised as one of the important landmarks in the analytical development of the lunar theory. In particular it was admired by Hill, whose own research was to provide

---

[27]The two primaries describe elliptic orbits, while the motion of the planetoid does not take place in the plane defined by the motion of the primaries.

[28]The mean anomaly is the angle the radius vector would have described if it had been moving uniformly with average speed $\dfrac{2\pi}{period}$.

[29]For a complete derivation of the equations see Hill [1876].

inspiration to Poincaré, and whose work is discussed at the end of this chapter,
and by Poincaré himself. In the introduction to Hill's collected works Poincaré
[1905] described Delaunay's use of canonical variables in perturbation theory as the
greatest contribution to celestial mechanics since Laplace.

2.3.7. Gyldén. 1881 saw the publication of the first of a long series of papers
by the Finnish astronomer and Director of the Observatory at Stockholm, Hugo
Gyldén. In these papers he developed his theory of absolute orbits for calculating
the motion of planetary bodies.[30] Gyldén's work, founded on the use of elliptic
functions, culminated in 1893 in the publication of the first volume of what was
intended to be a three-volume series devoted to the theory; the remaining volumes
were never completed due to his death in 1896.[31] Gyldén's system consists of the
sun and two planets, of which one planet is designated *disturbing* and the other
designated *disturbed*. The differential equations which represent the motion of the
disturbed planet are solved by means of sums of periodic terms whose arguments are
linear functions of a quantity which Gyldén called the planet's true longitude. The
terms that vanish when the disturbing mass is equal to zero are called *coordinated*
terms and are equivalent to the *periodic inequalities* of the classical theory. Those
that do not vanish when the disturbing mass is equal to zero but coalesce with those
that represent the elliptic motion of the disturbing planet around the sun are called
elementary terms and correspond to the *secular inequalities* of the classical theory.
Those that have the disturbing mass in the denominator of their coefficients are
called *hyper-elementary* (they do not occur in the final result), while those of long
period (which occur when the period of two planets are nearly commensurable) are
called *semi-elementary* or characteristic terms.

If, in the expression for the coordinates, all the coordinated terms are removed,
then the modified expression, which will only contain elementary terms, will define
a new orbit very close to the true one (the order of difference between them being of
the same magnitude as the disturbing forces). This new orbit is called the *absolute
orbit*. The solution of the differential equations is obtained by substituting expansions of the disturbing function into the differential equations and integrating. The
six arbitrary constants of integration are the elements that fix the absolute orbit of
the disturbed body.

Using the planet's longitude as the independent variable throughout the inte-
grations involves a large number of complicated transformations, and these, combined with the necessity of keeping the elementary and nonelementary terms separate, mean that the whole process is extremely complex. So much so that Hill,
when writing about the method, said:

> *A degree of complexity is thus imparted to the subject, which makes
> it difficult to see when one has really gathered up all the warp and
> woof of it. Professor Gyldén has nowhere removed the scaffolding
> from the front of his building and allowed us to see what archi-
> tectural beauty it may possess ... The advantages claimed for the
> method are that it prevents the time from appearing outside the
> trigonometrical functions, and that it escapes all criticism on the*

---

[30] For full references to Gyldén's papers see Whittaker [1899].

[31] A second volume of Gyldén's work on planetary theory, edited by Bäcklund with assistance
from Sundman and von Zeipel, was published in 1908. See Marcolongo [1919, 72].

score of convergence. The first is readily conceded, but many simpler methods possessing this advantage are already elaborated, and it is not clear that the second ought to be granted. No completely worked out example of the application of this method has yet been published. The great labor involved will naturally deter investigators from employing it.[32]

Furthermore, since Gyldén's constants of integration represented hitherto undefined quantities, an added difficulty in comprehension was provided by the new terminology which Gyldén invented to describe them. Nevertheless, Gyldén was not without his champions. One in particular was the German Martin Brendel,[33] who received his doctorate in 1890 for a thesis in support of Gyldén's methods. In a short paper [1889] on the three body problem, Brendel emphasised the practical significance of Gyldén's theory, although he did admit that it had suffered from misrepresentations.

With regard to the convergence of the series, in order to counteract the problem of small divisors (divisors of the order of the planetary masses), Gyldén modified the coefficient of the first power of the dependent variable by incorporating a function which he called the *horistic* (or limiting) function [1893]. But despite his efforts, which resulted in several lengthy papers, he did not in fact prove convergence, as Poincaré [1905a] was later to show.

However, prior to Gyldén's introduction of the horistic function in 1893, the question of the convergence of his series had already provoked contention between himself and Poincaré. In 1889 Gyldén, on learning of the content of Poincaré's Oscar prize-winning paper, had immediately claimed priority with a previously published paper of his own [1887]. The controversy, which was not limited to the main protagonists, is discussed in Chapter 6.

Nevertheless, in spite of their differences, Poincaré was a great admirer of Gyldén's methods, and openly acknowledged Gyldén's contribution to celestial mechanics on several occasions, not least in the introduction to both the first and second volumes of his *Méthodes Nouvelles*. Indeed he began the latter, which was an exposition of the work of contemporary dynamical astronomers, by proclaiming that *"it is the methods of Gyldén which are the best and to which I shall give the greatest exposure."*[34]

2.3.8. Lindstedt. The task of unravelling some of the obscurities in Gyldén's work seems to have provided the spur for the Swedish astronomer Anders Lindstedt to join in the search for trigonometric series solutions.[35] The first paper by Lindstedt to attract attention [1883] introduced a method for integrating an important class of differential equations that frequently occurs in perturbation theory in celestial mechanics. These equations, which essentially represent a perturbed harmonic oscillator, had arisen in Gyldén's researches, and were of the general form

$$\frac{d^2x}{dt^2} + n^2x = \alpha\Phi(x,t),$$

[32]Hill [1896a, *131*].

[33]Brendel made a distinguished career in mathematical astronomy, holding chairs at both Göttingen and Frankfurt.

[34]Poincaré [MN II, *v*].

[35]See von Zeipel [1921, *328*].

where α is very small and $\Phi(x, t)$ is a function expanded in powers of x with coefficients which are periodic functions of t. Subject to certain restrictions concerning the symmetry of the coefficients—Lindstedt had thought that the perturbing forces should be either odd or even functions of the angle variable involved—the method avoids secular and mixed secular terms and shows how the equations can be satisfied by x expanded as a trigonometric series.

Later the same year, Lindstedt applied the method in order to find trigonometric series solutions for the three body problem [1883a]. He began from the equations in Lagrange's 1772 paper. Making the assumption that the eccentricities, the ratio of the radius vectors, and the inclinations of two of the bodies were sufficiently small, he reduced the system to four second-order differential equations. He solved these by successive approximations, eventually eliminating all the secular terms so that the time only appeared in the arguments of the periodic functions. This gave him the coordinates of the three bodies as trigonometric series of four arguments, each of which was a linear function of time. He then (wrongly) assumed that it was possible to choose the constants of integration in such a way that convergence would be assured. Although, as will be described later, Poincaré exposed certain theoretical flaws in Lindstedt's method, from a practical point of view it was still very useful.

Lindstedt was not the first astronomer to provide such a series, chronologically the credit is due to Simon Newcomb [1874], who proved that the differential equations describing the motion of the planets could formally be satisfied by trigonometric series. However, since Lindstedt's method was the less complex of the two and consequently capable of greater generalisation, it became the more widely known. Lindstedt's method was also considerably simpler than Gyldén's, although it was correspondingly less powerful, as the increased simplification brought with it an accompanying reduction in the range of its application.

2.3.9. Hill. Shortly before Gyldén and Lindstedt started publishing in earnest, two papers appeared that were to influence deeply the future development of celestial mechanics in general and the three body problem in particular. In 1877 G. W. Hill privately published an exceptional paper [1877] on the motion of the lunar perigee. In it appeared the first new periodic solutions to the three body problem since Lagrange's discovery of special periodic solutions in 1772. In the following year, the first issue of the *American Journal of Mathematics* contained another important paper by Hill [1878] on the lunar theory, which included a more complete derivation of the periodic solutions. These two famous papers have long been acclaimed for the originality and elegance of the mathematical methods they contain, as well as for their substantial contribution to the progress of celestial mechanics. In particular, Hill's idea concerning periodic solutions had a profound effect on future research, most notably in the work of Poincaré.

One rather curious feature about these two papers is that they appear to have been published in reverse order. The paper of 1877 is plainly a logical continuation of the researches contained in the paper of 1878, and since Hill's work is characterised by brevity of expression, trying to understand the innovatory ideas in [1877] is especially difficult without the support of [1878]. Indeed such were the problems in comprehending some of Hill's work that Ernest Brown—the eminent theoretical astronomer who was one of Hill's strong supporters and later took up and continued his work, providing an almost exhaustive treatment of the lunar problem—was led to remark, *"Hill was not a great expositor; even for those familiar with the subject, his work is often difficult and sometimes obscure .. he is rarely anything else but concise."* [36]

Nevertheless, Hill's discovery and use of a new class of periodic solutions was a turning point in the history of the three body problem and dynamical systems in general. So original was Hill's approach that Poincaré, in the introduction to Volume I of his *Méthodes Nouvelles*, made the observation, *"In this work ... it is possible to see the germ of most of the progress that science has since made."* Whittaker was prompted to suggest that the publication of Hill's paper in 1877 signified *"the beginning of the new era of Dynamical Astronomy."* [37]

The prevailing influence on Hill's work came from Delaunay, whose two-volume work provided him with a major stimulus. He professed considerable admiration for Delaunay's method and, apart from employing it in the lunar theory, outlined ways in which it could be applied to other problems. Hill was also considered to be one of the few people who fully understood the work of another lunar theorist, the German Peter Hansen. Ranked as one of the leading theoretical astronomers of his day, Hansen's main work concerning the motion of the moon became the basis for extensive tables of lunar motion, published at the expense of the British Government in 1857.

The problem that concerned Hill in [1877] was the discrepancy between the theoretically computed values for the motion of the lunar perigee and those values derived from observation. Did this discrepancy arise because the approximations were not continued far enough, or was it because there were other forces acting on the moon which had not yet been considered? Since the question could only be answered if a limit could be put on the error incurred by the approximation method, Hill set out to compute the value of the motion of the lunar perigee *"so far as it depends on the mean motions of the sun and moon, with a degree of accuracy that shall leave nothing further to be desired."* [38]

Hill's innovation was to abandon the idea of using an elliptic orbit for the moon as a starting point, i.e., abandon the idea of neglecting the action of the sun as a first approximation, and instead begin with a circular orbit. He then used the effect of solar perturbation to vary the circular orbit before varying it again by the introduction of the eccentricity of the lunar orbit. In essence, he began by solving a modified version of the restricted three body problem before making a variation in order to attempt the general problem. Previous efforts had always begun by first solving the two body problem and then making the appropriate variation.

Hill had recognised that, of the five parameters involved in Delaunay's series for the longitude, latitude and parallax, the one whose expansion provided the slowest rate of convergence in the series was the ratio of the mean motions of the sun and the moon. This gave him the idea of beginning his attempt on the problem by neglecting the other lunar inequalities and finding the series in powers of this one alone. He embarked on this first stage of the problem in [1878], while he dealt with the second stage, which was to take account of the lunar eccentricity, in [1877]. He had originally intended to treat all five different classes of lunar inequalities as listed by Euler in [1772]—the two mentioned above, the lunar inclination, the solar eccentricity and the solar parallax—but shortly after [1877, 1878] had been

---

[36] Brown [1916, *293*].

[37] Whittaker [1899, *130*].

[38] Hill [1877, *243*].

published, Simon Newcomb, the director of the American Ephemeris, persuaded him to become involved in the theories of the motion of Jupiter and Saturn, and as a result he did not complete his original programme.

By initially only considering the ratio of the mean motions of the sun and the moon, Hill substantially simplified the differential equations. For, as pointed out by Poincaré [1905], by excluding the solar eccentricity and parallax, Hill had derived a formulation in which the sun could be said to describe a circle with a large radius. Hill's second insight was to choose rectangular coordinates uniformly rotating with the angular velocity of the sun, so that the time no longer appeared explicitly in the equations.[39] This choice of coordinate system was in contrast to the prevailing methods which invariably involved polar coordinates.

Using this formulation, the expressions for the coordinates of the moon referred to the rotating axes can be represented by Fourier series (with undetermined coefficients) and are periodic. The solution is obtained by substituting the Fourier series into the differential equations and determining the coefficients as functions of the parameter m, which depends on the ratio of the mean motions of the sun and moon. In order to avoid the multiplication of trigonometric functions and to enable a reduction to algebraic form, Hill took a further innovative step and introduced complex variables. Substituting the complex variables into the differential equations gave rise to an infinite system of algebraic equations from which the coefficients could be determined in terms of m, either algebraically or numerically. One particular advantage of Hill's method was the ease with which the approximation could be extended as far as desired.

Hill, having realised that the periodic solution he had found was of interest beyond its application to the lunar problem, varied the value of m by taking moons of 10, 9, ... , 3 lunations (the time from one new moon to another) in the periods of their primaries and obtained a family of different periodic solutions which he studied in detail. For moons of longer lunation, he found that the method was not practicable and resorted to mechanical quadrature. His final periodic solution involved a moon with a cusped orbit, which he called the moon of maximum lunation (*Figure 2.3.iii*).[40] However, this attribution was shown to be mistaken, first by Adams and then by Poincaré who subsequently proved that the cusp was succeeded by looped orbits (*Figure 2.3.iv*).[41]

Another fundamental idea that Hill introduced in [1878] was that of curves of zero velocity, which he used to show that the moon could never escape from its orbit around the earth.[42] This property is derived through the consideration of Jacobi's integral, which gives the square of the velocity relative to the moving axes. Since this quantity is necessarily positive, equating it to zero gives the equation of a surface which separates space into parts: those in which the velocity is real and those in which the velocity is imaginary. The surface consists of various curves and folds, and so is hard to understand in detail, but nevertheless it does reveal certain limitations on the motion of the moon's orbit. In particular it provides an upper limit for its radius, which allows certain conclusions to be drawn about the stability

---

[39] An idea for which Hill acknowledged his debt to Euler.

[40] Hill [1878, *331-335*].

[41] Hill [Collected Works I, *335*] and Poincaré [MN I, *109*].

[42] Coincidentally, Sylvester, who was the editor of *The American Journal of Mathematics* in which Hill's paper was published, had independently used a similar idea in an entirely different context. See Archibald [1936, *134*].

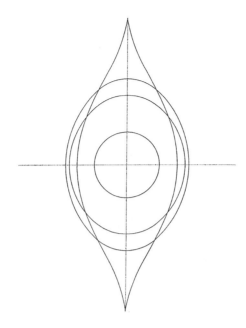

FIGURE 2.3.iii

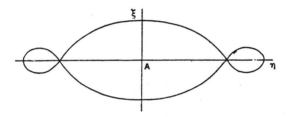

FIGURE 2.3.iv

of the motion. This was the idea taken up and used to great advantage by George Darwin [1897] (see Chapter 8).

In [1877] Hill first determined a periodic solution (now called the intermediate orbit or variational curve) and then derived the equations of variation, taking account of the first power of the lunar eccentricity. This led him to a fourth-order system of linear differential equations with periodic coefficients. By the use of known integrals combined with elegant transformations of his own he reduced the system to a single linear differential equation of second order [1877, *246*],

(7)
$$\frac{d^2 w}{dt^2} + \Theta w = 0,$$

where w is the normal deviation of the moon from the intermediate orbit and Θ only depends on the relative position of the moon with reference to the sun. Θ can, therefore, be expressed as a Fourier series

$$\Theta = \theta_0 + \theta_1 \cos 2t + \theta_2 \cos 4t + \cdots$$
$$= \sum_j \theta_j \zeta^{2j},$$

where $\zeta = e^{it}$ and $\theta_i = \theta_{-i}$, and so equation (7) can be written

$$(8) \qquad \frac{d^2 w}{dt^2} + w \sum_{j=-\infty}^{\infty} \theta_j \zeta^{2j} = 0,$$

which is now generally known as Hill's equation.[43]

If θ_0 is much greater than $\theta_i (i = 1, 2, \dots)$, then an approximate form of equation (7) is

$$\frac{d^2 w}{dt^2} + \theta_0 w = 0,$$

which has the particular solution

$$w = K\zeta^c + K'\zeta^{-c},$$

where K and K' are arbitrary constants and $c = \sqrt{\theta_0}$ is the ratio of the lunation to the anomalistic month.[44] When the additional terms of Θ are considered this has the effect of modifying the value of c and adding to w extra terms of the general form $A\zeta^{\pm c + 2i}$, and a particular solution of equation (7) is therefore

$$w = \sum_{i=-\infty}^{\infty} b_i \zeta^{c+2i},$$

where each b_i is a constant coefficient and the value of c is to be determined. Substituting this value of w into equation (8) and setting the coefficient of each power of ζ equal to zero generates a doubly infinite system of algebraic equations.

These equations can then be used to determine the ratios of all the coefficients b_j to one of them, b_0, which can be regarded as the arbitrary constant. If all the b_j are eliminated from these equations, then this gives rise to an infinite determinant involving c, denoted $\mathcal{D}(c)$, which, when equated to zero, determines c.

From this determinant Hill derived an expression for c in terms of the parameter m. In the first approximation, the value of c is $\sqrt{\theta_0}$, while in the second $c = \sqrt{1 + \sqrt{(\theta_0 - 1)^2 - \theta_1^2}}$.

This is a remarkably simple expression for an approximate value of the motion of the moon's perigee, and one which gives a value only about $1/60$ in excess of that given by observation, the difference being largely due to the neglect of the lunar inclination.

By performing further manipulations and transformations, Hill reduced the determinant to an infinite series in what he assumed to be a convergent form. Hill's final expression for the determinant turned out to be equivalent to starting with the equation $\mathcal{D}(c) = 0$, assuming $c = \sqrt{\theta_0}$ as the first approximation, and then expanding the expression $\sin^2\left(\frac{\pi}{2}c\right)$ in powers and products of the coefficients $(\theta_1, \theta_2, \dots)$. Taking this expression up to terms of order twelve in m, Hill achieved a value for c which was accurate up to the 15th decimal place. By contrast, he showed that Delaunay's method, calculated up as far as terms in m^9, produced a solution which was not even correct up to four significant figures, and, furthermore,

---

[43]Hill's equation can also be considered as a generalised form of Mathieu's equation

$$\frac{d^2 w}{dt^2} + (a + b\cos 2t)w = 0.$$

See Whittaker and Watson [1927, Chapter XIX].

[44]The time taken for the moon to pass from perigee to perigee.

he estimated that Delaunay's series would have to be prolonged up to terms in m^{27} to obtain a result of comparable accuracy to his own.

Immediately after the publication of Hill's 1877 paper, J. C. Adams, the English astronomer who had been among the first to predict the presence of the planet Neptune from unexplained perturbations in the orbit of Uranus,[45] communicated a brief paper to the Royal Astronomical Society [1877] in which he remarked that many years previously (1868) his own work on the lunar theory had followed a very similar course. Unfortunately his investigations had never been published, as he had considered them to be insufficiently complete. His method, which was rather less elegant than Hill's, dwelt on the motion of the moon's node and did not involve the use of rotating rectangular coordinates. As a result he been unable to find the rapidly converging series that Hill had deduced, although he had found the same infinite determinant.

Although the idea of an infinite determinant occurs in the literature prior to Hill's 1877 paper, it was not widely known or discussed, and Hill himself had not previously encountered the idea.[46] No formal theory existed, and one issue in particular gave him difficulties. Hill's results depended upon the convergence of the determinant, but it was a property he had been unable to prove.

The incompleteness of Hill's results meant that the idea of infinite determinants was not immediately taken up and utilised further. Almost another decade had to pass before the missing convergence proof was eventually supplied by Poincaré [1886d], and work on the topic was then continued by Helge von Koch. From then on the power of Hill's idea began to be recognised, and today Hill's position is secure as the founder of the theory of infinite systems of linear equations, his research in this field ranking alongside his contribution to dynamical astronomy as one of his principal achievements.

Despite the results he obtained in [1877, 1878], it was some time before Hill's work received the acknowledgement it deserved. In 1888, when Darwin began to study Hill's papers, he remarked that although they seemed to be very good scarcely anybody knew about them.[47] This could have been partly due to Hill's nationality—in the 1870s America was only just beginning to become established as a mathematical force—as well as his own rather solitary disposition. On the other hand there was the comparative inaccessibility of his particular style of mathematics as well as the deficiency in his theory of infinite determinants. Nevertheless, given the nature of Hill's research it is surprising that his work took so long to be appreciated.[48]

The breakthrough came in 1886. In that year not only did Poincaré prove the convergence of the infinite determinant but also Hill's 1877 paper was republished in *Acta* (the first paper in English to appear in the journal). And then in 1890, when the publication of Poincaré's memoir on the three body problem released

---

[45]Adams' predictions, originally made in 1845, were contemporaneous with those of the French astronomer Le Verrier and prompted a bitter priority dispute. The two astronomers took almost no part in the controversy, the feud being largely conducted by English and French journalists.

[46]According to Whittaker and Watson [1927, *36*] infinite determinants first appear in the researches of Fürstenau [1860].

[47]Brown [1916a, *lii*].

[48]For example, as Wintner observed [1941, *440*], the significance of Hill's periodic solutions escaped Heinrich Bruns, who reviewed Hill [1878] for the *Jahrbuch über dir Fortschritte der Mathematik* 10, *782*.

Poincaré's theory of periodic solutions, Hill's work on the progress of dynamical astronomy and the three body problem finally began to gain due recognition.

Poincaré clearly respected and appreciated Hill's abilities,[49] and there is no doubt that Hill was an important early influence on him, particularly in connection with his theory of periodic solutions. George Darwin even went so far as to suggest that Hill provided the inspiration for Poincaré to embark on his work in celestial mechanics.[50] And certainly Poincaré made no secret of his admiration for Hill. Sylvester, who was in Paris in January 1891, shortly after the publication of Poincaré's memoir on the three body problem, recorded that,

> *They speak great things here of Poincarré's* [sic] *prize memoir in the Acta—and he seems to have taken some of his most fruitful ideas from Hill of whom he speaks most highly both (as I noticed* [?] *in the memoir) and also in conversation as has been the case in talking with me.*[51]

Poincaré continued to maintain his interest in Hill's research, and when in 1904 he was in the United States for the St. Louis Congress and at last had the opportunity to meet Hill, he greeted him with the words, "You are the one man I came to America to see."[52]

---

[49]See Poincaré [1905].

[50]Darwin [1900, *517*]

[51]Letter to Simon Newcomb dated 8.1.1891. See Archibald [1936, *151-152*].

[52]Smith and Ginsburg [1934, *130-131*].

CHAPTER 3

Poincaré's Work before 1889

3.1. Introduction

Poincaré's memoir on the three body problem brought together a host of mathematical ideas and techniques that he had developed over the previous decade. Almost from the beginning of his academic career he had been concerned with the fundamental problems of celestial mechanics, and many of the papers he published during the 1880s relate to his interest in the subject. These include several of a broad theoretical nature as well as those in which he responded to explicit questions of dynamical astronomy.

First there is his acclaimed memoir on curves defined by differential equations. In this memoir, published in four parts between 1881 and 1886, Poincaré initiated the qualitative theory of differential equations in the real domain.[53] These papers are full of new ideas, many of which form the basis for results in [P2]. The three body problem features prominently, and Poincaré is quite clear about its motivating role in the development of his ideas.

Second, there are the papers in which he addressed either a particular aspect of the three body problem or a connected problem of celestial mechanics, some of which involved developing the work of another mathematician or astronomer. All of these papers are relatively short, none of them approaching the scale of [P2], but in them can be found his initial researches into periodic solutions and his early investigations into the convergence of trigonometric series used in celestial mechanics.

Finally, there are the papers in which Poincaré developed ideas and techniques which he used in [P2] but which were generated in a more general context. Notable amongst these are his thesis [1879] and his paper on asymptotic series [1886a].

3.2. The qualitative theory of differential equations

During the 18th century, the realisation that it was not possible to integrate the majority of differential equations using known functions had increasingly led to the study of the properties of the differential equations themselves.[54] By the middle of the 19th century this practice had become firmly established, and by the 1860s the success of complex function theory meant, with a few isolated exceptions, that the emphasis was firmly placed on the investigation of the behaviour of the function in the neighbourhood of a point in the plane. Thus at the beginning of the 1880s,

---

[53]Poincaré [1881], [1882], [1885] and [1886].
[54]See Kline [1972, Chapter 21].

when Poincaré began his memoir on the analysis of functions defined by differential equations, research was, in effect, centred on studying the local properties of a solution to a differential equation. Poincaré's approach was radically different. He looked beyond the confines of a local analysis and brought a global perspective to the problem, undertaking a qualitative study of the function in the whole plane.[55]

In [1880] Poincaré stated that his objective was to provide a *geometric* study of the solution curves of a first-order differential equation, and indeed it was his geometrical insight which became one of the the hallmarks of his work on differential equations. As Gilain [1991] and Gray [1992] have argued, what was new and important was Poincaré's idea of thinking of the solutions in terms of curves rather than functions, and it was this that marked a departure from the work of his predecessors, whose research had been dominated by power series methods.

Importantly, Poincaré's interest in differential equations was not driven only by an intrinsic interest in the equations themselves. He also had a particular interest in some of the fundamental questions of mechanics, most notably the question of the stability of the solar system, and he recognised the necessity of a qualitative theory of differential equations for furthering the understanding of this type of question. In this context therefore he saw it as important to consider the global properties of *real* as opposed to complex solutions. His attention to the real case marked another notable departure from the work of earlier investigators, which had been concentrated on the complex case.

Although the memoir was published as four papers, in terms of content it divides into two parts. The first two papers, which constitute the first nine chapters, were published in consecutive issues of Liouville's *Journal* in 1881 and 1882 and are devoted to the study of the simplest type of differential equation. The third and fourth papers, which appeared in 1885 and 1886, are concerned with equations of higher order and degree.

3.2.1. Papers I and II. Poincaré divided the study of a function into two parts: qualitative—the geometrical study of the curve defined by the differential equation—and quantitative—the numerical calculation of values of the function. At the beginning of the first paper he was quite explicit that his work was going to centre on the first part and that an element of his motivation for a qualitative study was his interest in the three body problem:

> *Moreover, this qualitative study has in itself an interest of the first order. Several very important questions of analysis and mechanics reduce to it. Take for example the three body problem: one can ask if one of the bodies will always remain within a certain region of the sky or even if it will move away indefinitely; if the distance between two bodies will infinitely increase or diminish, or even if it will remain within certain limits? Could one not ask a thousand questions of this type which would be resolved when one can construct qualitatively the*

---

[55] As Gilain [1991], in an excellent article, has pointed out, Poincaré's qualitative approach to differential equations was not entirely without precedent. In 1836 Charles Sturm made a qualitative study of second-order linear differential equations in the real domain, but his approach was set in a different context from that adopted by Poincaré. For a comparison between them see Gilain [1991, *224-225*]. Sturm's papers are treated in detail in Lützen [1990, *435-446*].

trajectories of the three bodies? And if one considers a greater number of bodies, what is the question of the invariability of the elements of the planets, if not a real question of qualitative geometry, since to show that the major axis has no secular variations shows that it constantly oscillates between certain limits.[56]

This point was later succinctly summarised by Jacques Hadamard when commenting on Poincaré's work:

The most important of them [questions of analysis and mechanics] *is well known, and its example presents itself with the whole spirit of the progress of astronomy: it is the stability of the solar system. The single fact that this question is essentially qualitative suffices to show the necessity of his* [Poincaré's] *point of view.*[57]

Apart from identifying the question of the stability of the solar system as an essentially qualitative problem, Poincaré had another stimulus for the qualitative consideration of differential equations: the analogy provided by research into algebraic equations. The proven success of qualitative investigations into algebraic equations supplied a clear indication of the potential of this approach.

Following the analogy, he began by constructing the curves defined by the differential equations. His initial researches centered on the simplest case: the construction of the solution curves of the equation

$$(9) \qquad \frac{dx}{X} = \frac{dy}{Y},$$

where X and Y are polynomials in x and y, and so $\frac{dy}{dx}$ is given as a single-valued rational function of x and y.

Although this equation, by virtue of its simplicity, has no direct application in celestial mechanics, by using it as the foundation for his study, Poincaré provided himself with a basis from which he could extend and elaborate his results to include more complex systems.

To circumvent the problem that was posed by the difficulty of the construction of curves with infinite branches, Poincaré first projected the plane onto a sphere.[58] The differential equation then associates a determined direction with each point on the sphere and no two solution curves can intercept except at singular points.

Looking for relationships between the different solution curves of the same differential equation, Poincaré began with a local analysis and examined the behaviour of these curves in the neighbourhood of a singular point. Unlike his predecessors Briot and Bouquet, who had studied singular points without the constraint of distinguishing between the real and complex case, Poincaré considered only real values. He showed that there were four possible different types of singular points and classified them by the behaviour of the nearby solution curves: *nœuds* (nodes), through

---

[56]Poincaré [1881, *4*].

[57]Hadamard [1912, *240*].

[58]The projection was made gnomonically, that is, the centre of the sphere is the centre of projection. Thus each point on the plane is projected into two points on the sphere and the projection of a straight line is a great circle.

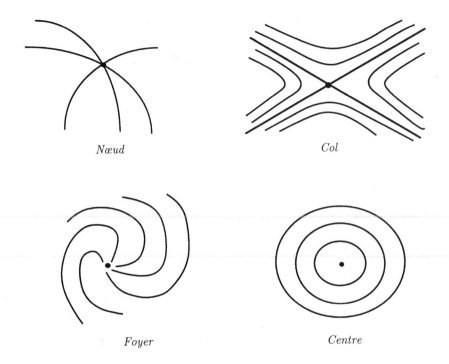

Nœud   *Col*

Foyer   *Centre*

FIGURE 3.2.i

which an infinite number of solution curves pass; *cols* (saddle points), through which only two solution curves pass, the two curves acting as asymptotes for neighbouring solution curves; *foyers* (foci), which the solution curves approach in the manner of a logarithmic spiral; and *centres* (centres), around which the solution curves are closed, enveloping one another (*Figure 3.2.i*).

Having used direct algebraic computation to show that these four types necessarily exist, he studied their distribution. He found that in the general case only three types prevailed—nodes, saddle points and foci—with centres arising only under very exceptional circumstances. This was quite unexpected, since earlier studies had shown that in the cases where elementary integration is possible the most usual singular points are nodes, saddle points and centres.

To describe the nature of a singular point Poincaré introduced the idea of an index to give a measure of the direction of the flow given by the solution curves about the singular point. Using this idea in relation to a system which could be described by equation (9)—that is for the case which corresponds to a simply connected surface with a single direction associated with each point on it, such as the flow on a sphere—he was led to a relationship between the different types of singular points: the number of nodes N plus the number of foci F was equal to the number of saddle points C plus two. This relationship, $N + F = C + 2$, is now known as the Poincaré index theorem for a flow on a sphere.

Poincaré next looked at the behaviour of solution curves beyond the neighbourhood of singular points. Here again the results he found were unexpected.

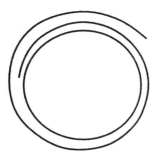

FIGURE 3.2.ii

Since the differential equation assigns a direction at each point of the solution curve, Poincaré took an algebraic curve on the surface and studied the direction of the solution curve at the points where the two curves cross. He called the points where the algebraic curve was tangent to the solution curve points of *contact*. He found that in a great number of cases there existed branches of closed curves, *cycles*, which were nowhere tangent to the solution curves, and these he called *cycles without contact*. Knowledge of the presence of these cycles is important because a solution curve cannot meet such a cycle in more than one point or there would be a point of contact, and so if it crosses such a cycle, it cannot recross it.

A second type of closed curve that played an important role in Poincaré's theory were those he called *limit cycles*. These are closed solution curves with no singular points but which are asymptotically approached by other solution curves (*Figure 3.2.ii*). The other solution curves spiral in towards a limit cycle but never actually reach it. Poincaré's discovery of their existence originated from his idea of indefinitely following a solution curve in one direction and looking at all the possible outcomes. Engaging in this process led him to the important result that every solution curve which does not end in a node is a cycle or a spiral. In other words, all solution curves are either closed or, with the exception of those which end in a singular point, asymptotically approach a limit cycle.

To prove the existence of limit cycles Poincaré considered the consecutive crossing points of a given (unclosed) solution curve C with an algebraic curve that cuts the solution curve in an infinite number of points.

To denote successive crossing points he introduced the term *consequents* (iterates) and said that if M_1 and M_2 were two consecutive crossing points, M_2 was the consequent of M_1 (*Figure 3.2.iii*). Then, using the fact that no two curves satisfying the equation can intersect, except at a singular point which is excluded, he showed that the successive crossing points must approach a common limiting position, say H. But since H is its own iterate, the solution curve through it is closed and is thus a limit cycle for the given curve C.

By obtaining results about the distribution of limit cycles, Poincaré began to generate a qualitative description of the flow described by the differential equation. He was able to determine the number of limit cycles within a given region of the sphere and also to find the particular regions in which a given number of limit cycles exist. In his next paper he gave a dynamical interpretation of these results.

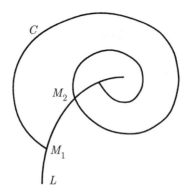

FIGURE 3.2.iii

3.2.2. Papers III and IV. Poincaré opened the third paper with a discussion on stability, which he began by reinforcing the connection with celestial mechanics:

> *One cannot read the first two parts of this memoir without being struck by the resemblance between the various questions which are treated there and the great astronomical problem of the stability of the solar system. This latter problem is, of course, much more complicated, since the differential equations of the motion of celestial bodies are of much higher order. Furthermore, one meets in this problem a new difficulty essentially different from those which we have had to overcome in the study of first order equations, and I intend to bring it out, if not in this third part, at least in the final part of this work.*[59]

This is the first occasion on which Poincaré explicitly tackled the question of stability, and in order to do so he began to use the language of dynamics to describe the differential equations. He reformulated equation (9) as

$$(10) \qquad\qquad \frac{dx}{dt} = X, \qquad \frac{dy}{dt} = Y,$$

where x and y are regarded as the coordinates of a moving point and t is the time. He said that the orbit of a point was stable if the point returned infinitely often arbitrarily close to its initial position.[60] Although this is essentially an impractical definition of stability in that it allows for intervening oscillations of any magnitude, nevertheless, from a theoretical point of view it does encompass a great degree of flexibility, and it allowed Poincaré (and later Birkhoff) to derive some remarkable results.

Poincaré first translated his results concerning closed curves into dynamical terms before considering them in the context of stability. In both cases, he showed that their presence is sufficient to indicate instability, although in the case of a limit cycle, it is the solution curves that asymptotically approach a limit cycle that are unstable, while the limit cycle itself is stable.

---

[59]Poincaré [1885, *90*]. Poincaré tackled the problem of higher order equations in the fourth paper, concentrating on the higher degree case in the third paper.

[60]A definition Poincaré later attributed to Poisson. See [P2, *313*].

Poincaré's analysis revealed that in the simple case described by equations (9) it was generally possible to cover the sphere with an infinite number of closed cycles of both types and, furthermore, that the number of limit cycles was usually finite. As a result, he concluded that in this case instability was the rule, stability the exception—the exception occurring when there were no cycles without contact and all the solution curves satisfying the differential equation were closed, mutually enveloping one another, as, for example, in the neighbourhood of a centre.

To study differential equations of first order but of higher degree, i.e., equations of the form

$$F\left(x, y, \frac{dx}{dy}\right) = 0,$$

F being a polynomial, Poincaré adopted a geometric representation. He considered the surface S with equation

$$F(x, y, z) = 0$$

and investigated the motion of a point upon it, putting the equations of motion into the form

$$\frac{dx}{dt} = X, \qquad \frac{dy}{dt} = Y, \qquad \frac{dz}{dt} = Z,$$

where X, Y, and Z are polynomials in x, y, and z such that

$$\frac{\partial F}{\partial x}X + \frac{\partial F}{\partial y}Y + \frac{\partial F}{\partial z}Z = 0.$$

In other words, he studied the solution curves of a vector field tangent to the surface S. In general, his discussion followed his treatment of the simpler case. He showed that the surface S was covered by an infinite number of closed curves which were either cycles without contact or limit cycles. He then examined the distribution of singular points and found that the relationship between them was given by

$$N + F - C = 2 - 2p,$$

where p is the genus of the surface S and is a fundamental invariant of the problem.[61] Since the sphere has genus 0, he recognised this as a generalisation of the relationship he had previously found for equations of first degree. Moreover, the relationship showed that the only surface upon which there can be a flow with no singular points is a surface with genus 1, i.e., a surface which can be identified with a torus. As Gray [1992, *511*] has observed, it was this result which led Poincaré to a detailed study of nonsingular flows on the torus, since, apart from a flow on a surface of genus 0, these are the simplest type of flows to study.

On the torus Poincaré chose variables ω and ϕ, with the differential equations given by

$$\frac{d\omega}{dt} = \Omega, \qquad \frac{d\phi}{dt} = \Phi,$$

where Ω and Φ are given continuous periodic functions of ω and ϕ which never vanish simultaneously.

In the particular case when

$$\frac{d\omega}{dt} = a, \qquad \frac{d\phi}{dt} = b,$$

---

[61]If at most $2p$ closed cycles can be drawn on a surface S without dividing the surface into two or more separate regions, then the surface is said to have genus p.

where a and b are positive constants, and the general solution is

$$\frac{\omega}{a} = \frac{\phi}{b} = constant,$$

Poincaré found that there were no limit cycles, and hence the orbit of the point was stable.

He next considered the cycle without contact defined by the meridian $\phi = 0$ and defined a set P to be the set of points M_i with $-\infty < i < +\infty$, where M_0 is the point on the meridian at time $t = 0$, M_i the ith iterate of M_0, and M_{-i} the ith antecedent of M_0. Then, skilfully employing Cantor's innovative idea of derived sets,[62] he was able to draw conclusions about the stability of the orbits of the points in the set P. Denoting P' as the derived set of P, he showed that either every point of P belongs to P' and the orbit of the point is stable or no point of P belongs to P' and the orbit is unstable; moreover, he showed that both these situations could occur. In other words, for equations with degree greater than one, stability was not the exception, in contrast to equations of first degree.

Poincaré also determined the circular order (in a fixed direction) in which the points of M were distributed on the meridian $\phi = 0$. Excluding the case of a limit cycle, he found that the order depended on an irrational number μ now known as the *rotation number* of the flow.[63] By defining α_i as the length of the arc $M_i M_{i+1}$, he proved that the limit as n tends to infinity of $\frac{\alpha_i + \cdots + \alpha_{i+n}}{n}$ is independent of i, i.e., it is independent of the initial point, and is equal to $\frac{2\pi r}{\mu}$, where r is the radius of the meridian. Having found μ he employed an elementary procedure to place each successive point M_i on the meridian. Then, using the fact that μ was irrational, he showed that the derived set P' was what Cantor called a perfect set.[64]

There was, however, one problem which Poincaré could not resolve. Given a differential equation where the set P' was the same for all orbits, he was unable to establish whether it was possible for certain orbits to belong to the case where every point of P belongs to P' while other orbits belonged to the case where no point of P belongs to P'. In other words, was it possible that some orbits could be stable while at the same time others were unstable? Or put in another way, was it possible for the perfect set P' to be disconnected? Although Poincaré was able to indicate some circumstances under which such a situation was impossible, he was unable to prove that it was always impossible.

He outlined how this problem was linked with the question of convergence of trigonometric series, in particular the method used by Lindstedt:

> *It is impossible not be struck by the analogy of this method of approximation with that of Lindstedt in celestial mechanics, and not to realise that the convergence of the process which I have shown is*

---

[62]Cantor [1872] defined the derived set of a given point set to be the set of its limit points. See Dauben [1979] for a full discussion of Cantor's work on set theory.

[63]Birkhoff [1920a, *87*].

[64]Cantor called a set perfect if it was unchanged by derivation. In other words, a set is perfect if every point in it is a limit point and the set is closed. See Dauben [1979].

*closely related to the convergence of series used by the learned as-
tronomer from Dorpat. But the problem that we have treated here is
evidently much simpler than the analogous questions of celestial mechanics, and if the difficulties are similar they are less numerous and
without doubt easier to overcome. It is this consideration which has
led me to persist with this question which has been the object of this
chapter and to which I shall doubtless return as I find new results.*[65]

To study higher order differential equations Poincaré considered X as the coordinates of a point moving in n-dimensional space with t as an auxiliary variable
representing time, so that the differential equations are written as

$$\frac{dx_i}{dt} = X_i, \qquad (i = 1, \dots, n),$$

where X are polynomials in x, and the solution curves are represented by the
trajectory of the point moving in n-dimensional space.

By considering the second-order case where the equations can be put in the
form

$$\frac{dx}{dt} = X, \qquad \frac{dy}{dt} = Y, \qquad \frac{dz}{dt} = Z,$$

he established the existence of five different types of singular points together with
three different types of singular lines defined by the points of Lindstedt's method
intersection of the three surfaces

$$X = 0, \qquad Y = 0, \qquad Z = 0.$$

He then classified the solution trajectories according to their behaviour in relation
to the singular points.

Poincaré also considered the question of finding solutions to the equations in
terms of convergent series valid for all real values of the time from $t = -\infty$ to
$t = +\infty$. In one of his earlier notes [1882a] he had considered the differential
equations defined by

$$\frac{dx_1}{X_1} = \frac{dx_2}{X_2} = \dots = \frac{dx_n}{X_n}.$$

He observed that if an auxiliary variable s is defined by equating the common value
of the above ratio to

$$\frac{ds}{X_1^2 + X_2^2 + \dots + X_n^2 + 1},$$

then it is possible to choose a positive constant α such that the x's can be expressed
by series in powers of

$$\frac{e^{\alpha s} - 1}{e^{\alpha s} + 1},$$

which are convergent for all real values of s, but only under the condition that the
trajectories do not meet a singular point except for infinite values of the variables
s and t.

He now considered this result in the context of the three body problem. He
found that for all real values of the time t the coordinates of the three bodies could

---

[65]Poincaré [1885, *157*].

be expressed in powers of

$$\frac{e^{\alpha t} - 1}{e^{\alpha t} + 1},$$

but only providing that it was known in advance that the distance between any two of the three bodies would remain greater than a given distance. In other words, the expansion is only valid providing there is never a collision between any two of the bodies. Since there is no way of foreseeing mathematically from the initial conditions whether or not such a collision will take place, this is a formidable gap in the theory. As a result, Poincaré was led to admit that he could not see any possibility of being able to take advantage of this method in celestial mechanics.[66]

With regard to using the distribution of the singular points as a way of gaining qualitative information about the behaviour of the solutions of the differential equations, Poincaré found that the increase in the number of different types of singular points made the investigation correspondingly more difficult. However, by returning to an idea that he had used in [1884], which involved a theorem due to Kronecker, and introducing Kronecker's index for a closed surface he was able to make some progress.[67] In [1884], details of which are discussed below, Poincaré had generalised Hill's idea of a periodic solution by applying Kronecker's theorem to the three body problem and thereby established the existence of an infinite number of periodic solutions. Now he considered the closed solution curves to the differential equations. What he found was that in the case of second-order equations the closed trajectories, which represent the periodic solutions, play an analogous role to that of the singular points in the first-order case. Thus in order to gain an understanding of the nature of the flow, instead of studying the solution curves close to a singular point he studied the trajectories neighbouring a closed trajectory. His idea of using periodic solutions in this way showed remarkable insight and one which was to be of central importance in [P2].

The next step in Poincaré's investigation is arguably one of the most important within the context of the paper, and one which proved fundamental both to [P2] and for the future of dynamical systems theory. He recognised that, rather than considering the flow in the entire three-dimensional space, it was much more convenient to consider the first return map induced by the flow on a two-dimensional surface of section transverse to the flow.

He therefore introduced the idea of a surface without contact (analogous to a cycle without contact), now better known as a transverse (or Poincaré) section. This is a surface $F(x, y, z) = 0$ on which there is no real point that satisfies

$$\frac{\partial F}{\partial x} X + \frac{\partial F}{\partial y} Y + \frac{\partial F}{\partial z} Z = 0,$$

and so F is not tangent to any trajectory. Again he considered the simplest type of solution space available within the constraints of the system: a torus without contact with no interior singular points.

He studied the different types of trajectory that approach a closed trajectory by examining their intersection with a transverse section. He chose axes in the

---

[66]Thirty years later Sundman utilised an analogous expansion with remarkable results. See Chapter 8.

[67]Kronecker [1869]. For indications of its history see Scholz [1980, *278-281*].

transverse section with an origin at the point where $t = 0$, and at each point M of the closed trajectory he constructed a transverse section. Since each point in the neighbourhood of the closed trajectory lies on a unique transverse section it can be represented by a point (x, y, t), where x, y are coordinates of the point in the plane of the transverse section and t represents the particular transverse section. Any trajectory starting infinitely close to the closed trajectory (say at m) will meet the transverse sections as the motion proceeds and will return to the original transverse section S when $t = 2\pi$, meeting it at the point m_1 (the iterate of m). The trajectory will continue and then meet S again at a third point m_2, the iterate of m_1, and so on. This is the idea of the first return map. In other words, the iterate of a point is just its image under the map. Thus in the three-dimensional space a periodic solution corresponds to a closed curve, but under the map a 2π periodic solution corresponds to a fixed point and a $2\pi k$ periodic solution corresponds to a cycle of period k. The problem of investigating trajectories is thus reduced to investigating the distribution of the successive iterates, or, in modern terminology, the iteration of a point transformation.

Providing the point transformation is sufficiently regular, it is possible, as a first approximation, to consider only its linear terms. Poincaré showed that the behaviour of the trajectories depended on the nature of the eigenvalues of the linear transformation and, furthermore, that a striking correspondence could be made between these eigenvalues and the different types of singular points. In [P2] he further investigated the properties of these eigenvalues (which he then named *characteristic exponents*) in order to deal with the question of stability of periodic solutions.

Studying the different situations corresponding to the different kinds of eigenvalue, he identified four types of trajectory: three whose behaviour could be easily understood, which he classified as corresponding to nodes, foci, and saddle points, and a fourth, which arose when the eigenvalues were conjugate pure imaginary, whose behaviour was more complex. Initially he thought this fourth type of trajectory corresponded to a centre, but it turned out to contain subtle and important differences. What he actually found was that if none of the constant terms in the trigonometric series that satisfied the differential equations were equal to zero, then this fourth case was equivalent to the first, and, moreover, this implied instability.

Poincaré next looked for the conditions under which all the constant terms in the trigonometric series were equal to zero. As he observed, this is the case which, although appearing extremely unlikely to occur, is the one that is encountered in the study of the general equations of dynamics. Importantly, he showed that the conditions under which the constants vanished involved another new and important idea, and one which was later to play a fundamental role in [P2], the idea of an *invariant integral*.

To understand what Poincaré meant by an invariant integral, consider a system of n first-order differential equations as defining the motion of a point in an n-dimensional space. Then consider a set of such points having a certain volume of space V at time t and so at any subsequent time t' the set has a new volume V'. If the volume V' remains constant whatever the value of t', then the volume is an invariant integral. For example, in the case of the differential equations of motion of an incompressible fluid, the volume is an invariant integral.

This was not Poincaré's first published reference to the concept. The idea first appears in [1886b], a paper published earlier in the same year. But in this earlier paper he had simply shown that it was the presence of a specific invariant integral that was the necessary condition for a particular result to hold, and he had not provided any discussion of the underlying theory—although, importantly, he had recognised that it was the presence of an invariant integral which accounted for the success of Lindstedt's method in eliminating the secular terms from the series used in celestial mechanics. It was knowledge of this property (unknown to Lindstedt) which, as he now explained, had enabled him to widen the class of equations for which Lindstedt's method was valid.

In his consideration of the question of the conditions under which the constant terms in the trigonometric series vanish, Poincaré first proved that the constant terms were equal to zero providing there was no closed surface without contact. He then showed that the existence of a certain invariant integral, which today is identified as the volume in phase space, was sufficient for this latter condition to be met.

Although at this stage Poincaré did not give any indication of the relationship between invariant integrals and the equations of dynamics, implicit in his argument is one of the important results of his theory of invariant integrals. For, as he later showed in [P2], Hamiltonian systems always admit the volume in phase space as an invariant integral. In other words, the existence of an invariant integral is a fundamental property of Hamiltonian systems, and hence the constant terms are bound to vanish. It was only in [P2], where the idea of an invariant integral played an essential role in his stability arguments, that Poincaré gave a name to the concept and developed a coherent theory.

Having discovered a condition under which stability was possible, Poincaré still had the problem of deciding when it occurred. He realised that, apart from the perennial problem of "small divisors" arising from the near commensurability of the frequencies of the interacting motions, the question also turned on whether or not the series concerned were uniformly convergent and under what conditions uniform convergence was assured. However, despite a long and detailed discussion of a particular example in which he examined the different situations that could arise depending on different initial conditions, he was forced to conclude that, even in the general case when the constants did all vanish, the series were not necessarily uniformly convergent. Hence the stability question was still unresolved.

To summarise, in looking at the behaviour of trajectories near a given periodic solution, what Poincaré had shown was that there were principally three different situations that could occur. First, the moving point could either continuously recede from or asymptotically approach the periodic solution, in which case the orbit of the point did not possess Poisson stability. Second, the moving point could oscillate within given limits close to the periodic solution, in which case the orbit did possess Poisson stability. Finally, the moving point could come arbitrarily close to any other point in the domain. In this case there was not only Poisson stability, since the point would always return as close as desired to its initial position, but there was also instability in the sense that the point could go arbitrarily far away from its initial position, and it was therefore impossible to assign limits to its coordinates. Since Poincaré's methods did not allow him to distinguish between the second and third situations, this presented a significant problem, not least because these two cases are precisely those which are encountered in the general equations of dynamics.

Thus, although he had established many substantial and important results for second-order differential equations, Poincaré's analysis was still less complete than that for the first-order case. Discussion of the periodic solutions had proved extremely fertile, but the difficulties of convergence and small divisors still remained. Poincaré ended the fourth and last paper in the series with his view of the implications of his results for the future progress of celestial mechanics:

> *From the above, one can easily understand the extent to which the problems due to small divisors and the quasi-commensurability of mean motions which occur in celestial mechanics result from the essential nature of things. It is extremely likely that these problems will arise whatever method is used.*[68]

3.3. Celestial mechanics and the three body problem

3.3.1. Trigonometric series. Poincaré's earliest work in celestial mechanics concerned the convergence of trigonometric series of the form

$$\sum A_n \sin \alpha_n t + \sum B_n \cos \alpha_n t,$$

which were used by astronomers to integrate differential equations such as

$$\frac{d^2 x}{dx^2} + n^2 x = \Phi(x, t).$$

These series are quite different from Fourier series in that the coefficients of time appearing as arguments inside the trigonometric functions are not proportional to integral coefficients and may decrease or increase indefinitely. The problem with the use of these series in celestial mechanics is in establishing their convergence. Are they convergent and, if so, is that convergence absolute? If they are absolutely convergent, are they uniformly convergent? If they are not absolutely convergent, are they what are now called asymptotic? These are all questions that had not been addressed until Poincaré made them the subject of a detailed mathematical investigation, most notably in [1884b].[69] In particular Poincaré was interested in the series derived by Gyldén and Lindstedt, and, beginning in 1882, he published several papers discussing the above issues, often specifically in response to the astronomers' work.

Poincaré's initial researches dwelt on the distinction between absolute and uniform convergence. His first result showed that if the convergence was not uniform then the function could attain arbitrarily large values, either by indefinitely increasing or by the amplitude of its oscillations indefinitely increasing [1882b, 1885a]. This was an important result in that it meant, contrary to what the astronomers had previously believed, that the ordinary convergence of a trigonometric series was not a sufficient condition for stability and so could not be used as a criteria for establishing results such as the stability of the solar system.

---

[68] Poincaré [1886, *222*].

[69] Hadamard [1922, *160*] described this paper as "... a work remarkable for its shortness and simplicity in comparison with its fundamental importance."

When Lindstedt [1883a, 1884] proposed a new series solution for the three body problem, Poincaré did not take long to respond [1883a, 1884b]. Lindstedt, instead of providing a formal proof of convergence for his series, had supposed that it was possible to choose the necessary constants in such a way that convergence was assured, at least for a given interval of time. Poincaré proved that if the series was absolutely convergent for such an interval of time, however short, then it was always convergent. He also showed that, since a function can only be represented by one such absolutely convergent series, there could not be two solutions to the problem. Furthermore, he pointed out that, although it was true that there were particular values of the constants for which the mutual distances of the three bodies could be expanded as convergent trigonometric series, it was by no means certain, indeed it was unlikely, that the convergence would subsist for other values of the constants, even for those arbitrarily close. This led him to the conjecture that Lindstedt's series were in fact asymptotic rather than absolutely convergent. That is, he thought that the series represented the mutual distances for a limited period of time only and that they did not do so indefinitely.

Later Poincaré was prompted by a paper of Gyldén's to consider the question of convergence in a slightly different context. Gyldén in his paper had been concerned with the problem of improving the convergence for a given trigonometric series. In [1886c] Poincaré showed that, providing the function and its derivatives were finite and they satisfied certain continuity conditions, then it was possible to find an upper bound for the coefficients in the trigonometric series representing the function and hence ascertain the strength of the convergence of the series.[70]

Poincaré's first published proof of the divergence of Lindstedt's series was contained in [P2], where it stood out as one of the results that he considered to be of particular importance. He later returned to the topic in the second volume of the *Méthodes Nouvelles*, where he gave a complete discussion of the methods used by astronomers in relation to the series used in celestial mechanics.[71]

In [1886b] Poincaré addressed a particular problem in Lindstedt's perturbation method. It will be recalled that Lindstedt's method involved a symmetry restriction which Lindstedt had introduced in order to ensure that at each stage of the approximation only one secular term was introduced, which was a necessary condition for his method to work. By a clever application of Green's theorem and, for the first time, using the idea of an invariant integral, Poincaré showed that the secular term which appeared in each approximation was necessarily unique, and therefore the restriction was unnecessary. In other words, the class of equations for which Lindstedt's method was valid was more general than Lindstedt himself had supposed.

In [1889] Poincaré gave a new derivation for Lindstedt's series using Hamilton-Jacobi theory. This new derivation had the advantage of completely bypassing

---

[70]Given a trigonometric series $\Sigma A_m \sin mx + \Sigma B_m \cos mx$, Poincaré said that the convergence of the series was of order p if

$$|m^p A_m| \leq K, \qquad |m^p B_m| \leq K,$$

where K is a positive quantity independent of m, and he then measured the strength of the convergence by its order.

[71]Details of this and Poincaré's later work in celestial mechanics are discussed in Chapter 7.

Lindstedt's symmetry restriction, which meant that his idea involving Green's theorem was no longer necessary. Four years later in [MN II] he went into this topic in greater detail.

3.3.2. Periodic solutions of the three body problem.

The best known of Poincaré's early papers in celestial mechanics is his first paper on periodic solutions of the three body problem [1884a]. It was published in the *Bulletin Astronomique* in 1884, an abstract [1883] having appeared in the *Comptes Rendus* during the previous year. It was this paper that he used as a basis for his investigation into the properties of closed solution curves in [1886] and, in turn, the theory he developed there provided the fundamental backdrop for his discussion on periodic solutions that formed the central part of [P2].

Poincaré's interest in the work of Hill and periodic solutions, a topic which dominated Poincaré's later researches in celestial mechanics, manifests itself here for the first time. By the application of a theorem due to Kronecker on solutions in systems of equations,[72] Poincaré proved that it was possible to choose the initial conditions for the three body problem in the case where two of the masses were very small relative to the third in such a way that the mutual distances of the three bodies were periodic functions of time. He thus proved the existence of a whole continuum of periodic solutions, thereby giving a generalisation of Hill's result.

Furthermore, Poincaré showed that this type of periodic solution could be classified into three different kinds:

1. Those in which the inclinations are zero, that is, all the bodies move in the same plane, and the eccentricities of the orbits are very small.

2. Those in which the inclinations are zero and the eccentricities finite.

3. Those in which the inclinations are finite and the eccentricities are very small.

He also speculated on the idea of a fourth kind of periodic solution in which both the inclinations and the eccentricities were finite, although he was unable to prove its existence except for certain values of the ratio of the two smaller masses.

Although the probability of the actual occurrence of such solutions was essentially zero (since they depended on particular values of the initial elements), Poincaré's insight was to realise that their importance lay in interpreting their relationship with other nearby solutions. He saw that if the initial elements of a solution were very close to those which corresponded to a periodic solution, it was possible to relate the true positions to the positions they would have occupied in the periodic solution and, to quote Gyldén, use this solution as an *intermediate orbit*. By supposing that the order of the inclinations and the eccentricities was sufficiently small so that their squares could be neglected, Poincaré showed that the differences between the true orbits and the intermediate orbits could be expressed by trigonometric series with no secular term. This greatly reduced the error that arose through the general method which involved the secular variation of the eccentricities.

---

[72]Kronecker [1869].

3.4. Other papers

Finally with regard to this survey of Poincaré's own precursors to [P2], there are two papers that are both related to the theory of differential equations. Although these are of a general nature and not specifically related to the three body problem, they both contain results which make an important contribution to the development of the theory in [P2].

3.4.1. Thesis.
Poincaré's thesis [1879], which was examined by Bouquet, Bonnet, and Darboux, concerned the study of integrals of first-order partial differential equations in the neighbourhood of a singular point.[73] It was his second paper on the theory of differential equations, and, as Hadamard remarked, it contained a strong pointer towards his future success with the topic and its applications to celestial mechanics: *"Even Poincaré's thesis contained a remarkable result which was destined later to provide him with a powerful lever in his researches in celestial mechanics."*[74]

Looked at in the context of the theory of differential equations already in existence at the time, Poincaré's thesis was the natural convergence of two streams of thought. On the one hand, Cauchy, and later Kovalevskaya, had applied Cauchy's method of majorants[75] to obtain results about the solutions of partial differential equations in the neighbourhood of an ordinary point, while on the other, Briot and Bouquet, and later Fuchs, using similar methods, had studied the solutions of ordinary differential equations in the neighbourhood of a singular point. Poincaré followed both Cauchy, by considering the solutions of partial differential equations, and Briot and Bouquet, by considering these solutions in the neighbourhood of a singular point.

Poincaré's analysis divided naturally into two parts according to whether the singularities under consideration were essential. In the case where the singularities were nonessential, he found that the solutions satisfied algebraic equations with coefficients analytic with respect to the variables. Although his treatment concerned a single partial differential equation, his results were analogous to those that would have been obtained by applying the theory to a system of ordinary differential equations, and it was in this analogous form that he applied the results in [P2].

With regard to the second and more difficult case concerning the essential singularities, one of the equations he considered was of the form

$$\frac{\partial z}{\partial x_1} X_1 + \cdots + \frac{\partial z}{\partial x_n} X_n = \lambda_1 z,$$

where the X_i can be expanded as powers of $x_1, \ldots, x_n$ with no constant term, and the first-degree terms can be reduced to $\lambda_1 x_1, \ldots, \lambda_n x_n$. He showed that this equation admits an analytic solution in $x_1, \ldots, x_n$ providing, first, there is no relation of the form $m_2 \lambda_2 + \cdots + m_n \lambda_n = \lambda_1$ where the m are positive integers, and,

---

[73]The examiners' report on Poincaré's thesis is reproduced in "Rapports sur les thèses des sociétaires (1870-1914) séries AJ16 Archives nationales", *Cahiers d'Histoire & de Philosophie des Sciences* **34**, 1991, *331*.

[74]Hadamard [1921, *206*].

[75]Cauchy called the method *calcul des limites* because it establishes the lower bounds or limits within which the series in question will necessarily converge.

second, that in the plane for the complex variable λ the convex polygon containing the points $\lambda_1, \ldots, \lambda_n$ does not contain the origin. This latter condition was the result to which Hadamard later referred, and its importance relates to the fact that it defines a space of nonexistence for the solutions of the equations. In [P2] Poincaré not only used this result explicitly but also further extended it for use in connection with his celebrated *asymptotic solutions*.

As in the previous case, there is a sense in which this single partial differential equation can be thought of as being equivalent to a system of ordinary differential equations, and seen in this light Poincaré's work can be considered as a generalisation of Briot and Bouquet's researches on a single differential equation.

3.4.2. Asymptotic series. With the emphasis on the rigorisation of analysis in the first half of the 19th century, the question of the legitimacy of divergent series became increasingly controversial. On the one hand divergent series were known to produce fallacious results if used indiscriminately, but on the other it was known that there were some divergent series, called semiconvergent series, which, for a given number of terms, provided an increasingly better approximation to the function as the variable increased—the best known example being that of Stirling's series for the gamma function.[76] In addition, it had long been recognised that some divergent series provided good numerical approximations for the functions they represented. In the latter part of the 19th century the increased application of these "useful" divergent series, particularly in dynamical astronomy, meant that there was a growing need to find some way of distinguishing them from other divergent series.[77]

In [1886a] Poincaré tackled exactly this question, and his solution provided the first formal definition of an asymptotic series. He began with a divergent series of the form

$$A_0 + \frac{A_1}{x^1} + \frac{A_2}{x^2} + \cdots,$$

in which he defined the sum of the first $n+1$ terms to be S_n. He said that a series of this type asymptotically represented a function $f(x)$ if the expression

$$x^n \left(f(x) - S_n \right)$$

went to zero as x increased indefinitely. In other words he had defined a general series which had exactly the same property as Stirling's series: the larger the value of the variable, the closer the series approximates the function. This can be put more formally, by saying that a series is an asymptotic expansion for a function $f(x)$ if for each n and each x sufficiently large but depending on n,

$$x^n \left| f(x) - S_n(x) \right| < \varepsilon,$$

---

[76]The gamma function is defined by: $\Gamma(z) = \int_0^\infty t^{z-1} e^{-t} dt$, for $Re(z) > 0$; and Stirling's series is given by:

$$\log \Gamma(z) = \left(z - \frac{1}{2} \right) \log z - z + \frac{1}{2} \log(2\pi) + \sum_{r=1}^{\infty} \frac{(-1)^{r-1} B_r}{2r(2r-1)z^{2r-1}}.$$

[77]The history of asymptotic solutions of differential equations in the 19th and early 20th centuries is well described by Schlissel [1976].

where ε is arbitrarily small. Thus the value of the function $f(x)$ can be calculated to a high degree of accuracy for large values of x by taking the appropriate number of terms in the partial sum S_n. Since the constants of the series are defined uniquely, it follows that if a function has an asymptotic series representation it is unique, although one asymptotic series can represent several different functions. Importantly, Poincaré showed that asymptotic series satisfy most of the same properties as convergent series, with the exception that in general they cannot be differentiated to form another asymptotic series.[78]

Having formalised the distinguishing property of these series, Poincaré applied the theory to a particular class of ordinary differential equations. Building on earlier results of Fuchs and Thomé,[79] Poincaré considered the integration of equations of the form

$$P_n \frac{d^n y}{dx^n} + \cdots + P_1 \frac{dy}{dx} + P_0 y = 0$$

where the P_i are polynomials in x.

If the equation has an irregular singular point at $x = \infty$, then Thomé had shown that in addition to m possible convergent series solutions $(m < n)$, there exist series of the form

$$e^Q x^a \left(A_0 + \frac{A_1}{x^1} + \frac{A_2}{x^2} + \cdots \right),$$

where Q is a polynomial in x, which formally satisfy the equation but which are generally divergent. It was these divergent series on which Poincaré focused his attention. Since the dominant characteristic of the series was the degree of the polynomial Q, Poincaré identified them by this property. Thus if the polynomial Q had degree p, then Poincaré called the series a normal series of order p.

First, Poincaré proved that the order of the differential equation's normal series solutions at $x = \infty$ could be determined. This involved introducing the idea of the rank of the differential equation, which he defined as follows.

Let M_i be the degree of the polynomial P_i and

$$N_i = \frac{M_i - M_n}{n - i}.$$

If h is the largest of the n quantities N_i, and k is the integer equal to or immediately larger than h, then Poincaré said that the equation has rank k at $x = \infty$, or the equation has a singularity of rank k at $x = \infty$. He then proved that if the differential equation has an irregular singularity of rank k at $x = \infty$, then all its normal series solutions at $x = \infty$ were of order k. Next, considering an equation with first-order normal series solutions, i.e., an equation of rank one, he proved that each series, although divergent, represented asymptotically one integral of the differential equation for large positive values of x. In the case of equations of higher rank, he first reduced them to rank one before attempting to find the asymptotic solutions.

---

[78]With regard to the differentiation of asymptotic series, consider the function $e^{-x} \sin(e^x)$.

[79]During the 1860s and 1870s Fuchs and Thomé had established important results concerning the solution of linear ordinary differential equations in the neighbourhood of a singular point. See Schlissel [1976] and Gray [1984].

Although Poincaré himself did not obtain many specific results using asymptotic series, the impact of the paper was far reaching. By creating a formal framework for these series, he provided a stimulus for investigations into asymptotic solutions of a variety of classes of ordinary differential equations. Moreover, from the point of view of his later work and in particular [P2], it will be seen that the theory played a fundamental role, most notably with regard to his discovery of asymptotic solutions of the restricted three body problem.

CHAPTER 4

Oscar II's 60th Birthday Competition

4.1. Introduction

In the autumn of 1890 Poincaré's memoir on the three body problem was published in the journal *Acta Mathematica* as the winning entry in the international prize competition sponsored by Oscar II, King of Sweden and Norway, to mark his 60th birthday on January 21, 1889.

A combination of royal patronage and carefully planned public relations meant that the competition achieved the unusual distinction of gaining recognition that stretched well beyond the world of mathematics. In the numerous obituary notices and commentaries on Poincaré's œuvre, not only is the memoir singled out for particular acclaim but the point is often made that it was as a consequence of winning the Oscar prize that Poincaré's name entered the public domain. The reference made by Paul Painlevé in a speech at Poincaré's funeral is characteristic of many:

> *In 1889, at the announcement of the result of the competition, France learnt with joy that the gold medal, the highest award of the new competition, had been awarded to a Frenchman, a young scholar aged thirty five, for a marvellous study of the stability of the solar system, and the name of Henri Poincaré became renowned.*[80]

Gaston Darboux, the Permanent Secretary of the French Academy of Sciences, expressed a similar view in a speech made at the Academy in honour of Poincaré:

> *From that moment on* [the announcement of the prize]*, the name of Henri Poincaré became known to the public, who then became accustomed to regarding our colleague no longer as a mathematician of particular promise, but as a great scholar of whom France has the right to be proud.*[81]

However, the paper which appeared in *Acta* differed remarkably from the version that had actually won the prize almost two years earlier. Its eventual publication drew to a close the competition, which, despite appearances to the contrary, had been beset with difficulties from its inception more than six years previously.

---

[80]P. Painlevé in Guist'hau et al, *Discours prononcées aux funérailles*, Gauthier-Villars, Paris, 1912.

[81]Darboux [1914].

Oscar II, King of Sweden and Norway

4.2. Organisation of the competition

By the late 19th century mathematical prize competitions had become well
established as a method for seeking solutions to specific mathematical problems.
These competitions usually emanated from the national academies, particularly
those in Paris and Berlin, the questions set reflecting the interests of the academy
concerned. Although the prizes offered were generally financial in nature, they
were valued much more in terms of academic prestige. Thus the existence of a
mathematical competition at this time was no novelty, but the Oscar competition
was somewhat unusual in that its sponsor, anxious that it should transcend national
barriers, did not associate his prize with an institution but chose rather to link it
to an academic journal, albeit one in which he had a personal interest.

Oscar was well known within mathematical circles; in her autobiography the
Russian mathematician and protégé of Karl Weierstrass, Sonya Kovalevskaya, who
spent the last years of her life as a professor of mathematics in Stockholm,[82] said
of him:

> King Oscar is a pleasant and cultivated person. As a young man he
> attended lectures at the university, and still today shows an interest
> in science, although I cannot vouch for the profundity of his erudi-
> tion. He has no official contact with the university but is extremely
> sympathetic to it and very amicably disposed towards its professors
> in general and to myself in particular.[83]

That Oscar should have sponsored such a competition was not altogether surprising. As a student at the University of Uppsala he had distinguished himself in
mathematics and from then on had become an active patron of the subject, providing financial support for various publishing enterprises, including the founding
of *Acta*, as well as making awards to individual mathematicians.

From its beginnings in 1884, the competition was organised by Gösta Mittag-
Leffler, professor of pure mathematics at the newly established Stockholm Högskola
(later the University of Stockholm) and founder and editor-in-chief of *Acta*.[84]
Mittag-Leffler had obtained his doctorate from the University of Uppsala in 1872,
and later had studied under Hermite in Paris, Ernst Schering in Göttingen, and
Weierstrass in Berlin. He thus had first-hand experience of life within the premier
mathematical communities in Europe, and this together with his involvement with
Acta meant that he was extremely well placed to promote the idea of an international competition.

Inspired by Weierstrass, Mittag-Leffler's own mathematical interests lay almost
entirely in the realms of analytic function theory. His *Habilitationsschrift* on the
foundations of the theory of elliptic functions had been published in 1876, followed

---

[82] At the time Kovalevskaya was the only female professor of mathematics in Europe. In the
years following her death in 1891 her reputation faded, but interest in her has recently revived.
For detailed accounts of her life and of her mathematics see Koblitz [1983] and Cooke [1984]
respectively.

[83] Kovalevskaya [1978, *228*].

[84] The first issue of *Acta Mathematica* appeared at the end of 1882. For the history of its
foundation see Domar [1982].

Gösta Mittag-Leffler

in 1877 by the first publication of the "Mittag-Leffler theorem" on the analytic representation of a single-valued complex function, and his later work focused on the problem of analytic continuation. Apart from being a talented mathematician he was also a skilled communicator. He assiduously cultivated and nurtured mathematical contacts at home and abroad, travelling extensively and maintaining a vigorous correspondence.[85]

Prior to the start of the competition, Mittag-Leffler had established a relationship with the King through the foundation of *Acta*,[86] but it is not clear whether the idea of holding the competition came from Mittag-Leffler or whether it can be attributed to the King himself. However, since Mittag-Leffler enthusiastically embraced any opportunity to raise the profile of Scandinavian mathematics, or indeed to enhance his own reputation within the mathematical community, and would certainly have relished the opportunity to be involved in an international mathematical competition, it seems probable that the project emerged as a consequence of his initiative.

What appears to be the first reference to the competition occurs in a long letter from Mittag-Leffler to Kovalevskaya written in June 1884, although its contents reveal that the topic was already under discussion. As the following extract outlining the proposed form of the competition shows, it is clear that from the outset the competition was intended to be one of pre-eminent importance in the field of mathematical analysis:

> *I agree with Weierstrass, if none of the answers on the set question are worthy of the prize, then the medal must be awarded to the mathematician who within recent years has made the best discoveries in higher analysis. ... we should not award our prize more frequently than every fourth year. Malmsten and the King want the prize jury to be appointed by the King and to consist of*
>
> *1. The main editor of Acta Mathematica*
> *2. A German or Austrian mathematician - = Weierstrass*
> *3. A French or Belgian mathematician - = Hermite*
> *4. An English or American mathematician - = Cayley? or Sylvester*
> *5. A Russian or Italian mathematician - = the first time Brioschi or Tschebychev, the second time Mrs Kovalevskaya.*
>
> *After each prize giving two of the prize judges should leave the jury and new ones should be appointed by King Oscar as long as he is alive—he must be able to appoint [substitutes] for both the leaving members. After King Oscar's death, the three remaining must appoint two new members but always in such a way as to fit the categories mentioned above.*[87]

---

[85]Mittag-Leffler's considerable correspondence, which is preserved at the Institut Mittag-Leffler, is described in Grattan-Guinness [1971].

[86]In 1882 Oscar had provided both financial and moral support to help Mittag-Leffler found *Acta*. See Domar [1982].

[87]Letter from Mittag-Leffler to Kovalevskaya 7.6.1884, I M-L (translated by S. Norgaard). For the complete extract see Appendix 1.

Charles Hermite

In fact, Mittag-Leffler was unable to fulfil any one of Oscar's requirements exactly. The difficulties with which he was faced are well illustrated by Kovalevskaya's reply, written while on holiday in Berlin:

> *In regard to the question of the prize Weierstrass has promised me that he will write you his opinion on that in more detail as soon as he receives a letter from you. I did not inform him of what you wrote me in the letter before last with regard to the choice of jury, for I was sure in advance of his complete disapproval. Indeed I believe that in this way the thing presents many practical difficulties. Just consider how one could hope that four famous mathematicians, Weierstrass, Hermite, Cayley and Tschebychev would <u>ever</u> agree on the merits of a memoir. I believe it is certain that each of the four would refuse to become part of the jury as soon as he learned the names of the other three. As for Weierstrass, I am so sure of this that I didn't even venture to talk to him about it. In general Weierstrass thinks that it will be quite difficult for the jury to agree when they have no opportunity to talk face to face. To do it by mail is considerably more difficult; and at bottom, why <u>would</u> these old gentlemen take so much trouble for us? There, I fear, is a very great difficulty! As for the honour, quite the contrary, each of the four that you named will be outraged that you chose the others along with him.*[88]

While there was an element of melodrama in Kovalevskaya's letter (Hermite and Weierstrass certainly had a healthy respect for each other), for the most part her presentiments proved to be well founded. The eventual outcome was a commission comprising only three: Hermite, Weierstrass, and Mittag-Leffler himself.

Hermite was one of the dominant figures of French analysis in the second half of the 19th century; from 1869 he was a professor both at the École Polytechnique and at the Sorbonne, resigning from the former in 1876 while maintaining his position at the latter until 1897. By the time of the competition he had established an international reputation in both teaching and research, and his courses attracted an audience from all over Europe. Not only was he a leading exponent of Cauchy's complex function theory, but also he actively promoted Weierstrass' ideas in France. His career had begun in the 1840s with work on elliptic and Abelian functions, topics which continued to occupy him throughout his mathematical life. By the late 1870s, having achieved notable success with his research into a variety of other topics, such as quadratic forms, invariant theory, and fifth degree equations, he returned once again to elliptic function theory. Throughout his life Hermite maintained an extensive and influential correspondence with other mathematicians, most notably with the Dutch mathematician Stieltjes,[89] but also, significantly, with Mittag-Leffler.[90]

---

[88]Extract from an undated letter (translated from French by R. Cooke). The letter is reproduced in full (in translation) in Cooke [1984, *106*].

[89]*Correspondence d'Hermite et de Stieltjes*, edited by B. Baillaud and H. Bourget, 2 volumes, Paris, 1905.

[90]The Hermite–Mittag-Leffler correspondence of 1874-1883; 1884-1891; and 1892-1900 is published in *Cahiers* **5** (1984), *49-285*; **6** (1985), *79-217*; **10** (1989), *1-82*, respectively.

Karl Weierstrass

In 1873 Mittag-Leffler had studied with Hermite in Paris, and a close friendship had developed between them. From the time of Mittag-Leffler's arrival, Hermite made no secret of the high regard in which he held the work of his German counterpart, Weierstrass. As Mittag-Leffler later recalled, his earliest memory of Hermite was of being greeted by the words, *"You have made a mistake, Monsieur, you should have taken the courses of Weierstrass in Berlin. He is the master of us all."*[91]

As a result of his connection with Hermite, Mittag-Leffler was able to remain in constant touch with the mathematical life in Paris. Moreover Hermite's friendship had proved to be extremely valuable with regard to the launching of *Acta*. Not only did Hermite show his support for the idea by sending him a handsome donation towards the initial financing of the project, but, more importantly, it was with Hermite's help that Mittag-Leffler had been able to secure, for the first issue of the journal, papers from three extremely talented young French mathematicians— Appell, Picard, and Poincaré—all of whom were Hermite's former students. And, importantly for Mittag-Leffler, all three of them continued to contribute to the journal, as did Hermite himself.

Weierstrass was a professor at the University of Berlin, a position he had held since 1864. He was a leading, if not the foremost, analyst in Germany, and his reputation as an expositor of new ideas drew students from across the world. Weierstrass had first come to prominence with his papers on Abelian functions published in 1854 and 1856,[92] having spent the previous decade establishing his theory of analytic functions on the foundation of power series. He lectured on a variety of topics, including several aspects of elliptic function theory as well as the theory of Abelian functions and the calculus of variations. He was also interested in the application of analysis to problems in mathematical physics, and in particular the n body problem.

However, Weierstrass' compulsion for rigour meant that he found it extremely difficult to complete anything for publication, with the result that his fame rested largely on his power as a teacher, and his influence was to a great extent carried by his former students, one of whom was Mittag-Leffler. While in Berlin Mittag-Leffler had established a good relationship with Weierstrass and, after he left in 1876, kept in regular contact through correspondence.[93]

Thus although Mittag-Leffler had failed in his original task of appointing a commission of five members, he had managed to engage two of the leading analysts of the day, one from each of the premier mathematical nations. Furthermore, they were two mathematicians with whom he had already established warm and productive friendships.[94]

---

[91]Mittag-Leffler [1902, *131*].

[92]"Zur theorie der Abelschen Functionen", *Crelle's Journal* **47** (1854), *289-306*; "Theorie der Abelschen Functionen", *Crelle's Journal* **52** (1856), *285-380*.

[93]Domar [1982] has noted that there was a slight lull in their correspondence at the beginning of the 1880s, which coincided with the founding of *Acta* and with Weierstrass being put in charge of *Crelle's Journal*.

[94]Not only did the composition of the commission not accord with the King's original conception, but also the idea of making the competition a regular event was never taken any further. That the competition was held only once was probably due both to the original difficulties in organising a commission and to the considerable problems that the commission later encountered.

However, despite the reduction in the number of people involved, such a choice of commission did still present certain practical difficulties. Apart from the obvious problems arising from the different geographical locations involved, Weierstrass in Berlin, Hermite in Paris, and Mittag-Leffler in Stockholm, there was the additional complication of the lack of a common first language. Although Mittag-Leffler was more than competent in both French and German and usually happy to use either, Hermite and Weierstrass, while familiar with each other's languages, each preferred to correspond in his own.[95]

The commission being appointed, Mittag-Leffler was faced with the formidable undertaking of achieving a consensus of opinion with regard to the question to be set. Naturally, for the competition to establish a international reputation, it was essential that it should attract entries of the highest international calibre which in turn would depend on the nature of the question to be solved. However, it soon became clear that to limit the competition to one question alone was going to be counterproductive. As pointed out by Hermite, there were by this time an unprecedented number of mathematicians working in different branches of analysis.[96] Thus to single out a particular topic on which to pose the question would be to impose a constraint that inevitably meant restricting the quality of the entries. Moreover, the imposition of such a limitation would preclude the inclusion of any work of an innovative nature. The difficulty was compounded by the fact that the King himself was eager for the competition to address a specific question. The possibility of having a single open question was discounted because of the fear that it might become impossible to choose a winner from several entries of comparable merit, each being on an entirely different topic.

An intensive correspondence ensued between all three members of the commission, with Hermite and Weierstrass exchanging ideas through Mittag-Leffler.[97] With the King becoming increasingly impatient, a format was finally agreed upon. The competition would consist of four questions and would give entrants the opportunity to submit an entry on a self-selected topic; however, priority would be given to entries attempting one of the nominated questions.

In mid-1885 the official announcement of the competition was published in both German and French in *Acta*,[98] as well as in English translation in *Nature*[99] and in several other languages elsewhere.[100] It gave details of the prize—a gold medal together with a sum of 2,500 crowns;[101] named the commission; listed the questions; and stipulated the conditions of entry.

---

[95]Mittag-Leffler's correspondence shows that he was extremely proficient in both French and German, but he did occasionally claim otherwise, as, for example, in a letter written to Kronecker in July 1885, which he began with, *"Please excuse me for writing to you in French. However badly I write French I find it easier and make less mistakes than in German."* Mittag-Leffler-Kronecker correspondence, I M-L.

[96]Hermite to Mittag-Leffler, 25.2.1885, No. 150, I M-L. *Cahiers* **6** (1985), 100.

[97]Mittag-Leffler used Kovalevskaya to translate Weierstrass' questions into French for Hermite. Letter from Mittag-Leffler to Hermite, 20.2.1885, No. 356, I M-L.

[98]*Acta* **7**, I-VI.

[99]*Nature* (30 July 1885), *302-303*. See Appendix 2.

[100]A list of the placings of the announcement is given in *Acta* **10**, 396.

[101]For comparison: Domar [1982] cites Mittag-Leffler's annual salary in 1882 as a professor in Stockholm as 7,000 crowns, and *Nature* (21 February 1889, *396*), in the announcement of the competition result, states that it is equivalent to £160.

Although Mittag-Leffler had originally suggested that Hermite and Weierstrass should each set two questions, it seems that, of the four questions set, the first three were proposed by Weierstrass and the fourth by Hermite. The first one addressed the well-known n body problem, reflecting Weierstrass' longstanding interest in the problem;[102] the second required a detailed analysis of Fuchs' theory of differential equations; the third asked for further investigation into the first-order nonlinear differential equations studied by Briot and Bouquet; and the last question concerned the study of algebraic relations connecting Poincaré's Fuchsian functions that have the same automorphism group.

The entries were to be sent to the chief editor of *Acta* before June 1, 1888, and, as was customary in such competitions, they were to be sent in anonymously, identifiable only by a motto and accompanied by a sealed envelope bearing the motto and containing the author's name and address. The entries were not to have been previously published, and notice was given that the winning entry would be published in *Acta*.

4.3. Kronecker's criticism

Unfortunately the announcement did not meet with universal approval. It provoked an angry reaction from another professor at the University of Berlin: Leopold Kronecker. Kronecker, apparently incensed at being left out of the commission, wrote to Mittag-Leffler with a catalogue of complaints.[103] But since it was no secret that an intense rivalry existed between himself and Weierstrass, it is very likely that he was more angry about Weierstrass' inclusion than he was about his own exclusion.[104]

Kronecker accused Mittag-Leffler of using the competition as a vehicle for advertising *Acta*. Why had the competition not been proposed by the Swedish Academy? It was an accusation Mittag-Leffler could easily refute: the King wished the competition to be announced in *Acta*, not only because *Acta* could claim a wider mathematical readership than the transactions of the Swedish Academy, but also because of his personal interest in the journal. On being challenged on the choice of members for the commission, Mittag-Leffler explained to Kronecker that his instructions had been to choose a commission of three, consisting of a representative from each of the top mathematical nations, Germany and France, with a third member from Sweden. With regard to the German representative, he told Kronecker that it had been a straight choice between him and Weierstrass, both being equally suitable, but Weierstrass, being some eight years the elder, had been chosen on the grounds of his "venerable" age. This may have mollified Kronecker,

---

[102]In a letter dated 15 August 1878, Weierstrass told Kovalevskaya that he had constructed a formal series expansion for solutions to the problem but was unable to prove convergence [Mittag-Leffler 1912, *31*], and in 1880/81 he gave a seminar on the problems of perturbation theory in astronomy [Moser 1973, *6*]. Despite Weierstrass' own difficulties with the problem, certain remarks made by Dirichlet in 1858 had led him to believe that a complete solution was possible, and hence his choice of the problem as one of the competition questions. Weierstrass' interest in the problem is chronicled in Mittag-Leffler [1912].

[103]The contents of Kronecker's letter to Mittag-Leffler have been reconstructed from Mittag-Leffler's reply written in July 1885, a copy of which is at the I M-L.

[104]Weierstrass believed that Kronecker's avowed antipathy to the work of George Cantor reflected Kronecker's opposition to his own work. See Biermann [1988, Chapter 5].

although it is doubtful whether Weierstrass would have been impressed by this line of reasoning.

However, Kronecker levelled his most serious charge at Question 4, the question set by Hermite. He maintained that he himself was the best person to judge algebraic questions of this type and that he had already proved that the results required to resolve this particular question were impossible to achieve; he threatened to tell the King as much.[105] As a defence, Mittag-Leffler could only plead ignorance on behalf of himself, Hermite and Weierstrass.[106] Mittag-Leffler concluded his reply with a barrage of flattery well calculated to appeal to Kronecker's vanity.

Kronecker then let matters rest, but not indefinitely. In 1888 he launched another assault; this time it was directed at the wording of Question 1. On this occasion he did not write to Mittag-Leffler but instead made his complaint at a meeting of the Berlin Academy of Sciences.[107]

Weierstrass, in composing the question, had drawn on information contained in a speech on Dirichlet given by Kummer.[108] This information had led Weierstrass to say that Dirichlet had told a "friend" that he had discovered a method for integrating the differential equations of mechanics and through this method had succeeded in proving the stability of the solar system. However, since Kronecker was the "friend" to whom Dirichlet had communicated his results, Kronecker felt he could claim to know what Dirichlet had really said and disputed the accuracy of Weierstrass' remarks. Kronecker's version of the events was that Dirichlet had first told him about the stability proof and then only later and on a separate occasion told him about the method; in other words, the two events were not contingent as Weierstrass had implied.

The content of Kronecker's second offensive would not have come as a complete surprise to Mittag-Leffler, since in August 1885 he had received a long letter from Kronecker, part of which centred on this question.[109] In addition, in October of the following year, 1886, Kronecker had openly declared that he considered Dirichlet to have been misquoted in the question and that he intended to publish his version of events.[110] Since Kronecker's complaint concerned unpublished work by Dirichlet, who had died in 1859, almost 30 years before, it is not clear why he waited a further three years before going public rather than pursuing the issue at the time he first raised it with Mittag-Leffler.

---

[105]The question made reference to solutions of equations of the 5th degree, and it appears that this was what Kronecker objected to. See the letter from Mittag-Leffler to Hermite, 8 August 1885, I M-L.

[106]Shortly afterwards Hermite met Kronecker and told him that he had proposed Question 4. He explained his intentions in setting the question, admitting that he had set it specifically with Poincaré in mind. Hermite's explanation seems to have satisfied Kronecker, as he did not pursue the issue. Perhaps it was sufficient that Weierstrass was not involved. See *Cahiers* **6** (1985), *108-111*.

[107]Kronecker [1888].

[108]See Appendix 2.

[109]Kronecker also disputed that the definition of "higher analysis" could be used to describe Questions 1 and 4. He also claimed to be highly competent to answer both these questions. Letter from Kronecker to Mittag-Leffler, 16.8.1885, I. M-L

[110]Mittag-Leffler to Hermite, 7.10.1886, Archives de l'Académie de Science, Paris. *Archive for History of Exact Sciences* **10** (1973), *162*.

Since this time Kronecker had chosen to make his views public, and the object of his censure was one of Weierstrass' questions, the attack appeared to be a further manifestation of the rivalry between the two Germans as opposed to a critique of the commission. Nevertheless, since Kronecker steadfastly maintained that he did not know who had composed the question, it was difficult for the commission to know how best to respond to him. Should they do so collectively and in the name of the commission, or should Weierstrass personally take on the responsibility?

Hermite made it quite plain that he did not wish to be involved in the dispute. Not only did he consider the matter to be an entirely German affair between the "two princes of analysis", but also he considered it his patriotic duty to avoid doing anything that could be construed as having a national connection.[111] He was convinced that Kronecker was a committed Francophobe and so felt there was nothing to be gained by his intervention.

Weierstrass made it clear that although he would have had no difficulty in dealing with Kronecker's complaints, he was reluctant to do so on his own since he considered the task a joint responsibility.[112] He believed that his own description of the events was essentially true. For even if Dirichlet had told Kronecker about the proof and the method at different times (which probably meant at the most one or two days apart), that did not mean that they had not been connected by Dirichlet. Likewise neither was the order in which Dirichlet related his discoveries to Kronecker evidence that that was the order in which he had discovered them. The only point Weierstrass was willing to concede was that he had omitted Kronecker's name as the "friend" to whom Dirichlet had communicated his results. In any case, from Weierstrass' point of view, what was important about Dirichlet's remarks was the fact that they provided real hope that a solution to the n body problem could be found and hence a good reason for including the question in the competition.

After much deliberation the commission decided against an immediate response in the belief that it would be better to wait until the judging of the competition had been completed and the winning paper(s) published. It turned out to be a wise decision. Not only did subsequent events overshadow the issue, but the need to reply was obviated by Kronecker's death in 1891, shortly after the publication of the winning memoirs.

4.4. The entries in the competition

Despite the fact that the identity of entrants for the competition was supposed to be secret, all three members of the commission were well aware that Poincaré meant to enter. As early as July 1887, Poincaré had made clear his intentions to Mittag-Leffler,[113] explicitly mentioning Question 1. In October of the same year Hermite told Mittag-Leffler that although he knew Poincaré was working on an entry for the competition, he did not know whether Poincaré would submit it; in any case, he was not sure whether Poincaré was working on Question 1 or Question 4. Mittag-Leffler, still scarred from Kronecker's original attack, admitted

---

[111]Hermite to Mittag-Leffler, 6.6.1888, I M-L. *Cahiers* **6** (1985), *140-142*.

[112]Weierstrass to Mittag-Leffler, 23.5.1888, I M-L. Mittag-Leffler [1912, *47-49*].

[113]Poincaré to Mittag-Leffler, 16.7.1887, I M-L. *Acta* **38** (1921), *162-163*.

to Poincaré that he hoped Poincaré would provide an answer to Question 4.[114] In fact the selection of topics for the competition was such that it would have been possible for Poincaré to have submitted an entry on any one of them. This raises the question: had they all been chosen with Poincaré in mind? Hermite freely admitted that this was the case with his question, and perhaps Weierstrass too had designed his questions to appeal particularly to Poincaré. Certainly Mittag-Leffler was an unquestionable champion of Poincaré's work.[115] In any event, Poincaré chose to attempt Question 1, the most difficult of the four.

By the closing date twelve entries had been received. Shortly afterwards a list of their titles, numbered in date order of submission, was published in *Acta*, with the authors identified solely by their respective mottos.[116] Five of the entrants, including Poincaré (number 9), had attempted Question 1, one had attempted Question 3 (number 4), and the remaining six had chosen their own topics.

When Poincaré's entry arrived it was clear that his reading of the regulations had been somewhat perfunctory. As required, he had inscribed his memoir with an epigraph,[117] but, instead of enclosing a sealed envelope containing his name, he had written and signed a covering letter and had also sent a personal note to Mittag-Leffler.[118] However, since he had already told Mittag-Leffler and Hermite of his intention to enter, and he knew that they would recognise his entry by its content— it was an explicit development of his earlier work on differential equations—as well as from his handwriting, it clearly was not a deliberate attempt to flout the procedures.

Apart from Poincaré, only the authors of entry numbers 4, 8, and 10 have so far been identified. With regard to entry number 8, the correspondence shows that the commission quickly established its author. The paper had been submitted with a covering note from Paul Appell, professor of rational mechanics at the Sorbonne and a regular contributor to *Acta*, claiming that it had been written by someone "well-known to him".[119] The commission originally surmised that the author was a student or friend of Appell's, but soon realised that it had been written by Appell himself.[120]

The authors of numbers 4 and 10 were identified as a result of their correspondence with the commission after the winner had been announced. Number 4 came from Guy de Longchamps, a professor in Paris, who, having a rather high opinion of his own work and having been passed over for the prize, saw fit to complain to Hermite (who did not share his opinion) about the manner in which the competition had been conducted.[121] Number 10 was the entry of Jean Escary, a professor at the Military School of La Flèche, who later became a professor at the Lycée de Constantine in Algeria. On learning of Poincaré's success he wrote to

---

[114]Mittag-Leffler to Poincaré, 17.11.1887, I M-L.

[115]Mittag-Leffler secured Poincaré's support for the launch of *Acta*, publishing important papers by him in each of the first five volumes.

[116]*Acta* **11**, *401-402*. See Appendix 3.

[117]*Nunquam præscriptos transibunt sidera fines* = Nothing exceeds the limits of the stars.

[118]Poincaré to Mittag-Leffler, 17.5.1888, Nos. 40, 41, I M-L.

[119]Appell to Mittag-Leffler, 13.5.1888, I M-L.

[120]Mittag-Leffler to Hermite, 17.10.1888, No. 1146, I M-L.

[121]Hermite to Mittag-Leffler, 4.2.1889. *Cahiers* **6** (1985), 160.

Mittag-Leffler enclosing some corrections to his own paper and praising Poincaré.[122] Earlier Mittag-Leffler, having spotted one of Escary's mistakes and ignorant of the paper's authorship, had confided in Weierstrass that he thought that the paper was by Dillner, a professor in Uppsala, since *"... for a long time the poor man has been unable to deal with mathematics."*[123]

Although there were officially twelve entries to the competition, the correspondence does reveal one further entry. This was an entry which was personally addressed to the King, but which arrived too late for consideration.[124] However, it is fair to say that had it arrived on time, it would not have added significantly to the commission's task, for the entrant, Cyrus Legg from Clapham, London, belonged to that indefatigable band of angle trisectors!

4.5. Judging the entries

A large part of the judging of the competition was done by correspondence. Mittag-Leffler, having received the entries in Stockholm, appointed one of the editors of *Acta*, Edvard Phragmén, the task of doing the preliminary reading before having copies of the most significant entries made and sent to Hermite and Weierstrass. Within a fortnight of the closing date Mittag-Leffler had written to both Hermite and Weierstrass with his opinion that there were only three entries worth considering, two from former students of Hermite—Poincaré and Appell—and one from Heidelberg,[125] although none of them had provided a complete solution to any of the given questions.

Mittag-Leffler spent August in Germany with Weierstrass so that they could study the memoirs together. The following month he wrote to Hermite to tell him that they thought Poincaré should win, with Appell being given an honourable mention. He made the point that Poincaré had the advantage inasmuch as he had at least attempted one of the set questions, whereas Appell had chosen his own topic: Poincaré had limited his investigations to a particular form of the three body problem (now known as the restricted three body problem) rather than the n body problem as specified in the question, while Appell had considered the expansion of Abelian functions by trigonometric series. Meanwhile Hermite had also been studying Poincaré's memoir and was equally convinced of the importance of the work.

The commission had quickly reached a unanimous decision, but the hard part of their work had not yet begun. It was one thing to recognise the quality of Poincaré's work but quite another to understand it. Poincaré's entry was not only extremely long (when printed for *Acta* it amounted to 158 pages), but it also contained many new ideas and results. Moreover, as Hermite freely admitted in a letter to Mittag-Leffler, the difficulties of comprehension were compounded by Poincaré's customary lack of detail:

---

[122]The *Jahrbuch über die Fortschritte der Mathematik* for 1889 and 1893 lists editions of Escary's paper as being published elsewhere but gives no further details.

[123]Mittag-Leffler to Weierstrass 16.11.1888, I M-L.

[124]Cyrus Legg to King Oscar II, 26.12.88, I M-L.

[125]Entry No. 5.

Edvard Phragmén

> *... But it must be acknowledged, in this work as in almost all his researches, Poincaré shows the way and gives the signs, but leaves much to be done to fill the gaps and complete his work. Picard has often asked him for enlightenment and explanations on very important points in his articles in the Comptes Rendus, without being able to obtain anything except the statement: "it is so, it is like that", so that he seems like a seer to whom truths appear in a bright light, but mostly to him alone . ...*[126]

All three members of the commission struggled with various parts of the memoir, but it was Mittag-Leffler who, determined that the version submitted to the King should be as complete as possible, entered into correspondence with Poincaré—notwithstanding the rules of the competition whereby he should have been ignorant of the memoir's authorship—appealing for clarification on several issues.[127] Poincaré responded by producing a series of *Notes* which, when printed, added a further 93 pages to the memoir.[128]

Mittag-Leffler may have had no qualms about his contact with Poincaré but Weierstrass was less happy about it and made a point of asking Mittag-Leffler not to mention the fact that he knew Poincaré had entered the competition.[129] He told Mittag-Leffler that it was almost an axiom in Germany for prize papers to be published exactly in the form in which they were submitted, although he personally considered the proper time for additions and corrections to be when the paper was being edited for publication, provided they were clearly acknowledged.[130]

Regarding the jury's opinion of the other papers, Weierstrass wrote to Mittag-Leffler in November with a report on five of the entries, although in reality the result of the competition had already been decided.[131] Of note in this report was his dismissal of number 5 as insufficiently mathematical; his recognition of the quality of Appell's paper, an opinion which had been further endorsed by Schwarz, to whom he had given Appell's paper to review; and, of course, his opinion on Poincaré's paper. He reiterated his belief that Poincaré's paper deserved the prize and singled out the particular results that he thought most important.[132]

4.6. The announcement of the result

Having worked through Poincaré's and Appell's memoirs, the commission now had to produce their reports. Weierstrass had the responsibility for writing the report on Poincaré's paper, Hermite for Appell's, and Mittag-Leffler for a general report on the competition. This apparently straightforward division of labour had not been without problems for Mittag-Leffler. Both Weierstrass and Hermite had shown reluctance for their tasks: Weierstrass because of his poor state of health,[133]

---

[126]Hermite to Mittag-Leffler, 22.10.1888. *Cahiers* **6** (1985), *147*.

[127]Mittag-Leffler to Poincaré, 15.11.1888, I M-L.

[128]For an indication of the nature and extent of these *Notes* see Appendix 5a.

[129]Weierstrass to Mittag-Leffler, 6.7.1888, I M-L.

[130]Weierstrass to Mittag-Leffler, 8.3.1890, 2.4.1890, I M-L.

[131]Weierstrass to Mittag-Leffler, 15.11.1888, I M-L. Mittag-Leffler [1912, *50-52*].

[132]Weierstrass' opinion of Poincaré's memoir is discussed in Chapter 6.

[133]Mittag-Leffler to Hermite 17.10.1888, No. 1146, I M-L.

and Hermite because of his connections with Appell—the two were friends as well as relations by marriage.[134]

On the 20th of January 1889, the day before the King's 60th birthday, Mittag-Leffler went to the palace (having belatedly obtained from Poincaré the necessary sealed envelope), and the result was officially approved. The general report was published in the newspaper *Postlidningen*, and Mittag-Leffler wrote to Poincaré to tell him that he would be receiving an official copy of the report via the Swedish ambassador in Paris.[135] Almost everything had been completed to the King's satisfaction; only Weierstrass' report on Poincaré's memoir was outstanding. Weierstrass, as Mittag-Leffler had feared, had been too unwell to fulfil his obligation within the allotted time, although he gave assurances that the report would soon be finished.

Needless to say, Kronecker too was keenly awaiting Weierstrass' report; as Mittag-Leffler confided to Poincaré, *"Kronecker is dreadful and is only waiting for the publication of the report so that he can criticise it."*[136]

Mittag-Leffler also wrote to the Academy of Sciences in Paris with details of the competition results, adding that the winning memoirs would be published in the next volume of *Acta*, due to appear in October.[137] The news of Poincaré's and Appell's success was well publicised in the French press,[138] and they were both made Knights of the Legion of Honour in recognition of their achievement. The French triumph also proved favourable for Mittag-Leffler, for he too was similarly honoured for his role in promoting French mathematics.

Unfortunately for Mittag-Leffler, the publication of the general report signalled the start of a distressing polemic between himself and the astronomer Hugo Gyldén, a fellow lecturer at the Stockholm Högskola and a member of the editorial board of *Acta*.

Although the report contained only a cursory indication of Poincaré's results, it was enough to convince Gyldén that he had anticipated Poincaré in a paper of his own [1887] published some two years earlier; he said as much in the February meeting of the Swedish Academy of Sciences. Once again Mittag-Leffler was placed in an uncomfortable position. Called upon by the King to defend Poincaré's memoir at the March meeting of the Academy, he wrote to Poincaré to explain his dilemma. Poincaré made a swift response, and Mittag-Leffler was able to complete the defence to his own satisfaction, although the issue continued to haunt him for several months.[139]

Mittag-Leffler's dispute with Gyldén paled into insignificance when compared with the problem that subsequently emerged. As already remarked, Mittag-Leffler, having allowed time for editing, had hoped to have the volume of *Acta* containing the winning memoirs published by October 1889. Apart from Weierstrass' report, for which he had continued to press, the actual printing was completed by the end

---

[134] Hermite to Mittag-Leffler, 22.10.1888. *Cahiers* **6** (1985), *147-149*.

[135] Mittag-Leffler to Poincaré, 21.1.1889, I M-L. For a copy of the report see Appendix 4.

[136] Mittag-Leffler to Poincaré 23.2.1889, I M-L.

[137] *Comptes Rendus* **108** (25.2.1889), *8*.

[138] Hermite to Mittag-Leffler, 28.1.1889. *Cahiers* **6** (1985), *159*.

[139] The controversy is explored in detail in Chapter 6.

of November. But the volume did not appear until the end of the following year, and when it did, it did not contain a replica of the memoir Poincaré had submitted to the competition.

4.7. Discovery of the error

The first glimmer that anything was awry occurred in July 1889. Phragmén, who was editing Poincaré's memoir for publication, alerted Mittag-Leffler to some passages in it which seemed to him a little obscure. Thus prompted, Mittag-Leffler wrote to Poincaré for yet further clarification.[140] However, it was not until much later that the scale of the problem became evident. Poincaré, in the course of dealing with Phragmén's queries, realised that he had made a serious error in a different part of the paper. At the beginning of December he wrote to Mittag-Leffler. Making no attempt to conceal his distress, Poincaré told Mittag-Leffler that he had written to Phragmén about the error, the consequences of which were more far-reaching than he had first thought. As a result Poincaré was making substantial changes to the memoir.[141]

This was most unwelcome news for Mittag-Leffler, since, although the volume of *Acta* had not been published, a limited number of printed copies of the memoir had already been circulated. Once more Mittag-Leffler's carefully nurtured mathematical reputation was in jeopardy. Despite his confidence in the overall quality of the memoir, he was only too conscious of the damage he personally would suffer should the scale of the error become public. Nevertheless, while total secrecy was impossible, he knew that if he could secure the return of the printed copies, then at least there would be no evidence to substantiate any rumours that might circulate. But while he knew to whom the copies had been sent, securing their return would not be easy, since several had been dispatched to foreign destinations. Apart from those sent to Hermite and Weierstrass, the recipients included Kovalevskaya, Jordan, von Dyck, Gyldén, Lindstedt, Lindquist, Lindelöf, Mohelins, Lie, Stone, and Hill.[142] Of those in circulation in Stockholm, Mittag-Leffler was particularly anxious to retrieve those in the possession of Gyldén and Lindstedt without arousing suspicion.

In a further attempt to minimise the scandal, Mittag-Leffler also suggested to Poincaré that everything concerning the error should be kept between themselves, at least until publication of the new memoir.[143] To safeguard himself still further, he gave detailed instructions to Poincaré about what he wanted in the contents of the introduction to the reworked memoir to ensure that no details of the error were included.[144] In addition he also asked Poincaré to pay for the printing of the original version—a request to which Poincaré agreed without demur, despite the

---

[140]Mittag-Leffler to Poincaré, 16.7.1889, I M-L.

[141]Poincaré to Mittag-Leffler, postmarked 1.12.1889. The contents of this letter are given in full at the end of 5.8.3.

[142]These names are listed by Phragmén in a note to Mittag-Leffler, 26.1.1890, I M-L. Despite the difficulties, Mittag-Leffler seems to have been tireless in his efforts to retrieve these pre-publication copies of the memoir. In the library of the Institut Mittag-Leffler there is one inscribed in Mittag-Leffler's hand with the phrase which in Swedish reads "whole edition destroyed".

[143]Mittag-Leffler to Poincaré, 5.12.1889, I M-L.

[144]Mittag-Leffler to Poincaré, 5.12.1889, I M-L.

fact that the bill came to just over 3,500 crowns, which was some 1,000 crowns more than the prize he had won.

Mittag-Leffler was also faced with the problem of telling Hermite and Weierstrass the unwelcome news about the error.

In Hermite's case, Mittag-Leffler had little choice but to be open, since he knew that Hermite would hear about it directly from Poincaré. He therefore wasted little time in sending Hermite copies of the correspondence between himself and Poincaré, adding that he thought the error so serious that it was likely most pages in the memoir would contain a false result.[145] Meanwhile, Hermite had seen Poincaré, who had told him about the error but said that it was not quite so bad as he had originally thought. Poincaré had offered to prepare a summary of his results for Hermite so that he could give a report to the Paris Academy of Sciences the following week.[146]

Unfortunately, Poincaré's original fears were realised. Exactly a week later Hermite was writing to Mittag-Leffler to tell him that he thought the situation was very serious after all.[147] He had heard nothing from Poincaré in the interim, the promised summary had never arrived, and he had not been able to make his report to the Academy. Not wishing to add further to Poincaré's distress, he decided to leave the matter entirely for Mittag-Leffler to handle.

As far as Weierstrass was concerned, Mittag-Leffler took an entirely different line. Initially he played down the seriousness of the problem and managed to give Weierstrass the impression that the delay in publication was due to the correction of some minor details. As a result, when, in February 1890, Gyldén and Wolf brought to Berlin rumours of serious errors in Poincaré's memoir, Weierstrass was placed in an extremely embarrassing position.[148] He was asked awkward questions that he was unable to answer. He demanded an explanation from Mittag-Leffler.

In defence, Mittag-Leffler claimed that his decision not to reveal everything about the error had been motivated purely by his consideration for Weierstrass' delicate state of health.[149] He told him that he believed Gyldén was acting in self-interest and that the situation was nowhere near so bad as Gyldén was trying to make out. Moreover, he claimed that the French mathematicians, including Poincaré and Hermite, were quite relaxed about the problem, and since the critics, such as Kronecker (whom he denounced as only being able to recognise something as important if he had done it himself) and Gyldén, were in a minority, there was no need for Weierstrass to worry! Weierstrass could do little except express his dissatisfaction at the way things had turned out and ask Mittag-Leffler to send him a proof of the new version as soon as possible.[150] He was clearly frustrated at not having discovered the error himself and also concerned about the potential inaccuracies in the general report, the extent of which he could not ascertain until he knew the details of the error. Fortunately, his worries in this direction turned out to be groundless. In the context of the revised memoir, the generalities contained in

[145]Mittag-Leffler to Hermite, 6.12.1889, No. 1372, I M-L.

[146]Hermite to Mittag-Leffler, 10.12.1889. *Cahiers* **6** (1985), *180-181*.

[147]Hermite to Mittag-Leffler, 17.12.1889. *Cahiers* **6** (1985), *181-182*.

[148]Weierstrass to Mittag-Leffler, 8.3.1890, I M-L.

[149]Mittag-Leffler to Weierstrass, 15.3.1890, I M-L.

[150]Weierstrass to Mittag-Leffler, 2.4.1890, I M-L.

the report still held true, and the lack of mathematical detail meant that everything in it could be applied to both versions.

Phragmén's role in setting Poincaré on the trail of an error that had escaped the attention of all three members of the commission was certainly worthy of recognition. However, and characteristically, Mittag-Leffler did not see it in his interests to make a public acknowledgement of Phragmén's participation. Nevertheless, he did ask Poincaré for written support to help Phragmén in his attempt to secure the chair in mechanics at the University of Stockholm,[151] and Phragmén was promoted to the editorial board of *Acta* in the following year. In November that year Phragmén himself revealed his interest in Poincaré's memoir by publishing a paper [1889] in which he showed that some of Poincaré's results could be applied to dynamical problems other than the restricted three body problem.

4.8. Publication of the winning entries

By the beginning of January 1890 Poincaré had completed his reworking of the memoir and sent a copy to Phragmén for editing. Not only had he made substantial alterations to accommodate the corrections arising from the error but also, where appropriate, he had incorporated the explanatory *Notes* into the paper itself. Thus in two quite distinct ways the memoir took on a significantly different appearance from that of its predecessor.

Although printing began at the end of April that year, a backlog of other work meant that it was not completed until the middle of November. When Volume 13 of *Acta* eventually appeared, it contained both Poincaré's and Appell's memoirs, together with Hermite's report on the latter.

Weierstrass's report on Poincaré's memoir was promised for a future volume but was never completed. Before the discovery of the error, Weierstrass had got as far as writing the introduction and, in March 1889, had sent it to Mittag-Leffler. However, the introduction was concerned only with general issues connected with the question as set and made no specific references to Poincaré's paper. So although the comments in it were not invalidated by the error, it did not provide the much-needed guide to Poincaré's paper. Nevertheless, the introduction was certainly not without interest, and Mittag-Leffler [1912] selected it to appear in his biography of Weierstrass, which was published in *Acta* and which focused solely on Weierstrass's interest in the n body problem.[152]

Given Mittag-Leffler's initial concern over obtaining Weierstrass's report, it might seem somewhat surprising that he was not able to induce him to complete it. However, after the discovery of the error, there was a marked reduction in Mittag-Leffler's interest in the report. Weierstrass had made it quite plain that he felt a moral obligation to make public the history of the error, but Mittag-Leffler's preoccupation with his own reputation meant that he was extremely anxious to play down the error's importance and was therefore keen for Weierstrass to do likewise.

---

[151]Mittag-Leffler to Poincaré, 4.12.1889, I M-L.

[152]Mittag-Leffler [1912, *63-65*]. A discussion of this introduction is given in Chapter 6, where it is put into context with Weierstrass's private remarks about Poincaré's memoir, many of which Mittag-Leffler also saw fit to publish in [1912].

It is tempting to assume that Mittag-Leffler considered it in his own best interests for Weierstrass's report never to appear.

Thus, over a year later than Mittag-Leffler had originally planned, the climax to the competition—the publication of the winning entries in *Acta*—finally took place. More than six years had elapsed since Mittag-Leffler had written optimistically to Kovalevskaya with the original plans for the competition. Despite Kovalevskaya's foreboding, Mittag-Leffler could scarcely have foreseen the turbulent course of events that was to follow. Nevertheless, in the final analysis Mittag-Leffler's considerable efforts were rewarded. Once the *Acta* volume was in circulation, the rumours of the error faded and the brilliance of Poincaré's work was freely acknowledged. Mittag-Leffler's hope that the competition would result in some important new mathematics had been amply fulfilled. Poincaré's memoir had ensured that King Oscar's 60th birthday celebration would not be forgotten.

CHAPTER 5

Poincaré's Memoir on the Three Body Problem

5.1. Introduction

Poincaré's attack on the three body problem was remarkable in many ways. His unprecedented qualitative approach to the problem and its intrinsic dynamics is unequivocally more powerful than any previous methodology. By taking a reductionist view and studying the periodic solutions of a system with two degrees of freedom, Poincaré's global qualitative perspective led him to the brilliant discovery of asymptotic solutions. His analysis of the complex nature of the behaviour of these solutions marks a turning point in the history of dynamics. For attendant on this analysis came his discovery of homoclinic points, which embodied the first mathematical description of chaotic motion in a dynamical system. Furthermore, as the comparison of the two versions of the memoir makes clear, Poincaré's first encounter with homoclinic points, as well as being historically important in its own right, has an added significance in the context of the history of the paper itself.

Furthermore, many of the innovative and powerful ideas that Poincaré developed as tools and techniques in order to tackle the three body problem have a more general application not only in the theory of differential equations and celestial mechanics but also in other branches of mathematics. The memoir therefore fulfilled a research role well beyond the confines of the problem it professed to tackle.

For Poincaré himself, the competition had acted as a stimulus to synthesise many of the ideas that he had been developing over the previous decade. The question of the stability of the solar system was one in which he had harboured an interest for several years, and he had for some time been building up a battery of techniques with which to launch an offensive. With the memoir completed, undoubtedly more hurriedly than he would have liked, he continued to work further on these ideas. Three years later saw the publication of the first volume of his three-volume *chef d'œuvre* on celestial mechanics, *Les Méthodes Nouvelles de la Mécanique Céleste*,[153] which was a work founded upon the contents of [P2].

The previous chapter described the circumstances that resulted in Poincaré's making two major changes to the content and structure of the memoir before its publication. The first of these, the addition of the substantial explanatory *Notes*, was made in response to requests for more detail from Mittag-Leffler, and the second, the extensive rewriting, was the result of the discovery of the error. Although Poincaré touched on the subject of these changes in his introduction to the published paper [P2], he did not make clear the extent of the alterations. Unfortunately, it has not been possible to trace the paper Poincaré originally submitted for the

---

[153]Henceforth referred to as [MN, I-III].

prize, but correspondence at the Institut Mittag-Leffler suggests that, excluding the *Notes*, it assumed a very similar form to the first printed version [P1], copies of which still exist at the Institut. Of especial importance amongst the latter is the one, [P1a], which was personally corrected and extended by Poincaré and to which he attached a note detailing the changes. This copy in its altered form corresponds almost exactly to the published memoir, and it provides a remarkable record of the way the memoir was rewritten. Its existence allows us to follow the metamorphosis of the entire memoir and to provide a complete picture of the exact nature of the error.

This chapter, as well as giving a detailed mathematical analysis of the memoir, also describes how [P2] relates both to the version that actually won the prize and to the *Notes*. The comparison of the versions shows how much of [P1] was retained in [P2], how the *Notes* were integrated (or not) into [P2], and to what extent [P2] was shaped by the detection of the error. Chronicling Poincaré's changes in this way gives a clear picture of why the discovery and correction of the error is all-important with regard to the position the memoir now holds in the history of the mathematical theory of chaos.

Comparing the tables of contents of both printed versions gives a preliminary idea of the overall structure of each version and its relationship to the others.[154] (The differences detailed here are also noted at the appropriate points in the mathematical analysis.) The two introductions are also compared, as this gives an insight into how Poincaré's own perspective on his results changed.

5.2. Tables of contents

Both versions of the memoir are prefaced with an introduction, then separated into two parts: *Generalities* and *The equations of dynamics and the n body problem*. The first part is devoted to developing the theory and the second to applying it. Each part is divided into chapters which are then subdivided into sections. The only difference in the format is that in [P1] the sections are numbered within each chapter, while in [P2] the section numbering runs straight through the memoir, which makes it easier for cross-referencing. [P1] concludes with nine explanatory *Notes*, each topic being labelled with a letter *A* through *I*.

Comparing the tables of contents shows two major changes in the first part of the memoir. Chapter I in [P2] contains four sections as against one in [P1]. In addition to the identical first section on notation and definitions, [P2] includes two sections, §2 and §3, on the method of majorants and a section, §4, on the integration of linear differential equations with periodic coefficients. Most of §2 and all of §3 are taken from *Note E*, and §4 is an exact reproduction of *Note D*. Although the format of Chapter II is identical in both versions, there are significant changes to the content. *Note C* is incorporated at the end of §6, and §8 contains major alterations. Chapter III of Part 1 contains the most important change, with the addition in [P2] of a new concluding section, §14, on the asymptotic solutions of the Hamiltonian equations; the contents of this, apart from including the latter half of

---

[154]The two tables of contents are reproduced as Appendices 5a and 5b respectively.

Note I, do not appear in [P1]. The other sections in Chapter III, although carrying the same headings in both versions, contain significant changes and additions.

In Part 2 the differences are more marked.

In [P1] the application of the theory to systems with two degrees of freedom is confined to the first chapter, which is divided into five sections. The second chapter is devoted to a general resumé of the results (positive and negative), and the final chapter consists of a single section on Poincaré's endeavours to generalise his results to the *n* body problem. [P1] concludes with the nine *Notes*.

In [P2] the first section corresponds with that in [P1], but it is the only section in the first chapter. Significantly, the topic of asymptotic surfaces has been revised to merit a chapter in its own right. The new second chapter consists of four sections, none of which retains an exact title from [P1]. One of the sections, §17, contains material from [P1]; one, §18, contains an amended version of *Note F*, together with some additions; and two, §16 and §19, are entirely new. The third chapter, on miscellaneous results, includes as §20 the section on periodic solutions of the second kind taken from the first chapter in Part 2 of [P1]; a section on the divergence of Lindstedt's series, §21, taken both from the negative results in [P1] and *Note A*; and a section on the nonexistence of uniform integrals, §22, which contains the rewritten contents of *Note G*. The last chapter in [P1] on the *n* body problem is transferred almost intact as the last chapter of [P2]. The section on positive results from [P1] is omitted altogether.

As far as the *Notes* are concerned, with the exception of *Note B, New statement of results*, which was entirely deleted, and *Note H, Characteristic exponents* (rewritten and expanded to appear as part of §12), these are incorporated, either whole or in part, into the main text of [P2] as indicated above. The exclusion of *Note B*, which was a summary of the main results described in more practical terminology for the benefit of astronomers, is discussed later.[154a]

Whenever a particular piece of the memoir is not specifically ascribed to either [P1] or [P2], it is correct to assume that it appeared in the same form in both versions.

5.3. Poincaré's introductions

In [P1] Poincaré began the introduction by admitting that, although he had written the memoir in response to Question 1 in the competition, he had not been able to provide a full resolution of the problem.[155] He made it clear that he had concentrated on the restricted problem which he specified as follows:

> *I consider three masses, the first very large, the second small but finite, the third infinitely small; I assume that the first two each describe a circle around their common centre of gravity and that the third moves in the plane of these circles. An example would be*

[154a]See Chapter 5, end of Section 5.7.
[155]For details of the Question see Appendix 2.

*the case of a small planet perturbed by Jupiter, if the eccentricity
of Jupiter and the inclination of the orbits are disregarded.*[156]

He then gave an indication of the main mathematical techniques that he had
used in the memoir. These included the trigonometric form of the power series
solutions derived by Lindstedt and Gyldén (which he had used to avoid the secular
terms that arise in the series used by Laplace and Poisson); Cauchy's method of
majorants, which he had applied to prove the convergence of the series; his own
geometric methods (taken from his earlier memoir on differential equations), which
he had used to prove the stability of the solution; and his new idea of invariant
integrals, the theory of which he had developed in order to apply his geometric
methods to the equations of dynamics.

He emphasised that the central topic of the memoir would be provided by his
discussion of periodic solutions and drew attention to the fact that he had been
able to develop the theory using Cauchy's methods, since the periodic solutions
were untroubled by the problem of small divisors.

[P1] was printed as though it were an exact replica of Poincaré's competition
entry and so retained its "anonymous" format. However, even without indicators
such as handwriting, a cursory reading of the introduction would have been sufficient to identify the author, since in order to furnish the necessary background to
his methods Poincaré needed to reference his own work, and consequently give his
own name, no less than five times.

In the case of [P2] Poincaré began by revealing that it was a reworking of his
competition entry. He explained that the revision had resulted from incorporating
the *Notes*, together with some additional explanations, into the main body of the
memoir, a task he considered to be a logical necessity but that he had not had time
to do earlier. Although he did mention the error, even acknowledging Phragmén's
role in its detection, he adhered to Mittag-Leffler's request and gave no hint of what
it might have been.

Nevertheless, he did make it clear that he had included some substantial additions to the opening chapter by the way of reformulation of established theorems.
In drawing attention to his work on periodic solutions, he mentioned both asymptotic and doubly asymptotic solutions and in this connection indicated the nature
of the restricted three body problem, although this time without giving a complete
statement of the problem. He also mentioned his recurrence theorem, but above
all he stressed what he called his negative results. These were his proof of the
nonexistence of any new single-valued integrals for the restricted problem and his
proof of the divergence of Lindstedt's series. To give an indication of the difficulties
he had encountered in attempting to generalise his results, he said that he believed
a complete solution to the three body problem would require analytic tools quite
different and infinitely more complicated than any of those known, which was in
contradiction to what Weierstrass had supposed in setting the question.

Finally, in connection with one of the series he had discussed, Poincaré acknowledged an analogy with a paper by Karl Bohlin [1888] which had been published so

---

[156]Poincaré [P1, *8*].

near to the closing date of the competition that he had not included a reference to it in his memoir.

Comparing the two introductions reveals quite a striking change of emphasis. Poincaré's focus is no longer on mathematical techniques but instead is on so-called negative results. [P1] conveys a sense of optimism about the ultimate resolution of the problem, but in [P2] the tenor is quite different: the future progress of the problem has lost its air of inevitability. The differences between them provide a fair reflection of the far-reaching and totally unexpected mathematical implications of the memoir's essential revision.

5.4. General properties of differential equations

In the first chapter, Poincaré provided the definitions and background for the theory to follow. Although most of the terminology he used would have been familiar to his contemporaries, the fact that the memoir was directed towards an international audience meant there was a special need for precision to avoid ambiguity or misunderstanding.

He considered the system of ordinary differential equations

$$\frac{dx_i}{dt} = X_i \qquad (i = 1, \dots, n),$$

where the X_i are single-valued analytic functions of the n variables $x_1, \dots, x_n$. Given a particular solution $x = \phi(t)$, and a nearby solution $x = \phi + \xi$, he defined the *variational equations*

$$\frac{d\xi_i}{dt} = \frac{dX_i}{dx_1}\xi_1 + \cdots + \frac{dX_i}{dx_n}\xi_n.$$

He called the system *canonical* if the variables x could be divided into two series $x_1, \dots, x_p$, and $y_1, \dots, y_p$, where $n = 2p$, and the equations could be written

$$\frac{dx_i}{dt} = \frac{\partial F}{\partial y_i} \qquad \frac{dy_i}{dt} = -\frac{\partial F}{\partial x_i} \quad (i = 1, \dots, p).$$

For the case when $n = 3$ he made a geometric representation of the system using his idea from [1885] and [1886]. That is, the x_i are considered as the coordinates of a point P in space so that as the time varies P describes a trajectory. By considering the set of trajectories which pass through a given curve in space, he formed a surface which he termed a *surface trajectory*. This representation then led to a definition of stability in which he said the system was stable if all its surface trajectories were closed. In other words the system was stable if any point P stayed within a bounded region of space.

To define a *periodic solution* he said that a particular solution,

$$x_1 = \phi_1(t), \dots, x_n = \phi_n(t),$$

of the differential equations was periodic with period h if when x_i is a linear variable

$$\phi_i(t + h) = \phi_i(t)$$

and when x_i is an angular variable

$$\phi_i(t + h) = \phi_i(t) + 2k\pi \quad (k \text{ an integer}).$$

Having dealt with the definitions, Poincaré moved on to establishing some of the techniques he intended to employ.

In broad terms the three body problem is essentially one of integrating a particular system of differential equations using series methods. While the problem's persistent resistance to attack had made it seem probable that some radical new ideas would be needed if progress was to be made, Weierstrass, in his formulation of the problem for the competition, had indicated that he thought it likely that the answer lay within the bounds of current analytical theory. Poincaré's success, which came not from solving the problem per se but in providing a whole new perspective on it, was the result of a complementary mixture of both old and new ideas and, in particular, the ingenious application of them. He built on a framework of existing results, enhancing and extending their utility in novel and inventive ways.

To launch the theoretical part of the memoir he reviewed some of the available techniques, beginning with the method of majorants.

5.4.1. The method of majorants. The method of majorants originated with Cauchy in [1842] in the search for proofs for the existence of solutions to differential equations. Generally speaking, the method is used to show that a power series in the independent variable (derived by the method of undetermined coefficients) that satisfies the differential equation does have a definite domain of convergence. It had been simplified by Briot and Bouquet [1854], used by Weierstrass in [1842] (although not published until 1894),[157] studied by Fuchs,[158] and, as mentioned in Chapter 3, Poincaré himself had worked on it in his thesis published in 1879.

Poincaré presented Cauchy's basic principle in the following form:

Given a system of differential equations

(11) $$\frac{dy}{dx} = f_1(x, y, z), \quad \frac{dz}{dx} = f_2(x, y, z),$$

where f_1 and f_2 can be expanded in increasing powers of x, y and z, then the equations have a unique solution

$$y = \phi_1(x), \quad z = \phi_2(x),$$

where ϕ_i are Taylor series in x which vanish with x.

To verify that such a solution exists, the series must be shown to be convergent. The two functions f_1 and f_2 are replaced by the majorant function

$$f(x, y, z) = \frac{M}{(1 - \alpha x)(1 - \beta y)(1 - \gamma z)},$$

M, α, β, γ being chosen in such a way that each term of f has a larger coefficient (in absolute value) than the corresponding term in f_1 and f_2. Replacing f_1 and f_2 by f increases the coefficients of ϕ_1 and ϕ_2, and since the series for f is convergent, the two series created by the exchange must be convergent, which in turn implies convergence of the original series for f_1 and f_2.

---

[157]Weierstrass's work became known to his students and colleagues in the late 1850s. See Cooke [1984, *28*].

[158]See Gray [1984].

In [P1], Poincaré made extensive reference to Cauchy's results, but his own exposition of the method as well as some further developments which he had derived were contained only in *Note E*. In [P2] he extended these developments and added two new sections, §2 and §3, to deal with theory as applied to ordinary differential equations and partial differential equations, respectively. These sections contain almost all the results from *Note E* as well as three new theorems, Theorems III, V and VI.

These new theorems are particularly important within the context of the memoir, since Poincaré's theory of periodic solutions depends fundamentally upon them. By including them in [P2] Poincaré put into place an essential foundation of the theory. In [P1] he used the results frequently but often with little or no reference, which made it extremely difficult to validate his arguments. Although not connected with the error, the addition of these sections represents a significant contribution towards his aim of creating a more logical structure to the memoir.

Most of the theorems contained in §2 are now well known, but they are stated here for completeness.

In the first theorem, Poincaré extended Cauchy's original result by finding an expansion for the solution in terms of a parameter μ as well as in terms of the independent variable t.

THEOREM I. *Suppose that the functions f_1 and f_2 depend not only on x, y and z but also on an arbitrary parameter μ and that they can be expanded as series in x, y, z and μ. Then equations (11) can be written in the form*

$$(12) \quad \frac{dx}{dt} = f(x, y, z, \mu) = 1, \quad \frac{dy}{dt} = f_1(x, y, z, \mu), \quad \frac{dz}{dt} = f_2(x, y, z, \mu),$$

and it is possible to find three series

$$x = \phi(t, \mu, x_0, y_0, z_0) = t + x_0, \quad y = \phi_1(t, \mu, x_0, y_0, z_0), \quad z = \phi_2(t, \mu, x_0, y_0, z_0)$$

that formally satisfy the equations. These series reduce to x_0, y_0 and z_0, respectively, for $t = 0$, and, provided t, μ, x_0, y_0 and z_0 are sufficiently small, they are convergent.

In his proof Poincaré replaced the functions f, f_1 and f_2 by the function

$$f'(x, y, z, \mu) = \frac{M}{(1 - \beta\mu)(1 - \alpha(x + y + z))}$$

and formed majorant series for x, y and z convergent for sufficiently small values of t, μ, x_0, y_0 and z_0. However, although this gave the desired result—that the series solution is an expansion in ascending powers of the parameter as well as the independent variable—it also contained a severe restriction on the value of t. Since ultimately Poincaré was looking for solutions valid for all t, he wanted to find some way of relaxing this restriction. In the following theorem, which has become a classic in the theory of differential equations depending upon a parameter, he showed that it could be done by proving the existence of a solution which is an expansion in powers of the parameter rather than powers of the independent variable.

THEOREM II. *Excluding one exceptional case, x, y and z can be expanded as powers of μ, x_0, y_0 and z_0 for any value of t, provided μ, x_0, y_0 and z_0 are sufficiently small.*

In his proof Poincaré began by showing that equations (12) have a solution

$$x = \omega_1(t, \mu), \quad y = \omega_2(t, \mu), \quad z = \omega_3(t, \mu),$$

which is such that $x = y = z = 0$ when $t = 0$ and which converges when $0 < t < t_1$. He then replaced x, y, and z in the equations (12) by

$$x = \xi + \omega_1, \quad y = \eta + \omega_2, \quad z = \zeta + \omega_3$$

to get the variational equations

$$(13) \qquad \frac{d\xi}{dt} = \phi(\xi, \eta, \zeta, \mu), \quad \frac{d\eta}{dt} = \phi_1(\xi, \eta, \zeta, \mu), \quad \frac{d\zeta}{dt} = \phi_2(\xi, \eta, \zeta, \mu),$$

ϕ, ϕ_1 and ϕ_2 vanishing when $\xi = \eta = \zeta = \mu = 0$. Since he had supposed that f, f_1 and f_2 can be expanded in powers of μ, then the same will be true of ϕ, ϕ_1 and ϕ_2, and these expansions can also be shown to be convergent in $0 < t < t_1$. Thus there exists a solution of equations (13) as series in μ, which is such that $\xi = \eta = \zeta = 0$ when $t = 0$ and which converges in $0 < t < t_1$.

The exceptional case occurs when the functions f_1 and f_2 are no longer analytic in the variables x, y and z, i.e., when they become infinite or cease to be single-valued. This is because if the functions are not analytic then they cannot be expanded in series as required. In other words if as t changes the trajectory goes through a singular point, the theorem no longer holds. In the case of the three body problem the functions given by the equations cease to be analytic in the case of a collision. However, since Weierstrass had specifically excluded collisions in the competition question, the theorem was sufficient for Poincaré's purpose.[159]

In his next theorem Poincaré proved explicitly that the solutions depend analytically on the initial conditions. This theorem did not appear in [P1], and it seems likely that he originally believed the result to be self-evident from Theorem II.

THEOREM III. *Let*

$$x = \omega_1(t, \mu, x_0, y_0, z_0), \quad y = \omega_2(t, \mu, x_0, y_0, z_0), \quad z = \omega_3(t, \mu, x_0, y_0, z_0)$$

be the solutions of the differential equations which reduce to x_0, y_0, z_0 for $t = 0$. Then the functions

$$\omega_i(t_1 + \tau, \mu, x_0, y_0, z_0), \quad (i = 1, 2, 3),$$

can be expanded as powers of μ, x_0, y_0, z_0, and τ, provided that these quantities are sufficiently small.

Poincaré attributed Theorem IV, now more familiarly known as the implicit function theorem, to Cauchy and his method of majorants.[160] Although it was not a new result, he had included it in *Note E* because it played such a pivotal role in his own investigations. Theorems V and VI, which only appear in [P2], are direct extensions of it.

---

[159]Poincaré appears not to have considered the possibility of noncollision singularities. The impossibility of such singularities in the three body problem was proved by Painlevé in [1897]. See Chapter 8.

[160]The history of the implicit function theorem is convoluted and worth further research. It is certainly not clear that Poincaré was right in his attribution.

THEOREM IV. *A system of* n *equations*

$$f_i(y_1, \ldots, y_n, x_1, \ldots, x_p) = 0, \qquad (i = 1, \ldots, n),$$

where the f *are analytic functions of the* $n + p$ *variables* y *and* x, *and vanish with them, can be solved for* $y_1, \ldots, y_n$ *in increasing powers of* $x_1, \ldots, x_p$, *if the Jacobian of the functions* f *with respect to* y *is not zero when* x *and* y *vanish together.*

The final two theorems and the accompanying corollaries take account of the case when the Jacobian does vanish. Poincaré did not include them in [P1], although he did make use of the results. In [P2] he did not provide proofs but instead referred to his own thesis and to work by Pusieux.

THEOREM V. *Let* y *be a function of* x *defined by the equation*

$$f(y, x) = 0,$$

where f *can be expanded in powers of* x *and* y. *Suppose that for* $x = y = 0$, f *and* $\dfrac{df}{dy}, \dfrac{d^2 f}{dy^2}, \ldots, \dfrac{d^{m-1} f}{dy^{m-1}}$ *vanish, but* $\dfrac{d^m f}{dy^m}$ *does not vanish. There will exist* m *series of the following form*

$$y = a_1 x^{1/n} + a_2 x^{2/n} + \cdots$$

(n *a positive integer,* a_1, a_2 ... *constant coefficients*) *which satisfy the original equation.*

COROLLARY I. *If the above series satisfy the equation, then so does the series*

$$y = a_1 \alpha x^{1/n} + a_2 \alpha^2 x^{2/n} + \cdots,$$

where α *is an* nth *root of unity.*

COROLLARY II. *The number of series of the form given in Theorem V which can be expanded in powers of* $x^{1/n}$ *(which cannot be expanded in powers of* $x^{1/p}, p < n$) *is divisible by* n.

COROLLARY III. *If* $k_1 n_1$ *is the number of the series which can be expanded as powers of* x^{1/n_1}, *and if* $k_p n_p$ *is the number of the series which can be expanded as powers of* x^{1/n_p}, *then*

$$k_1 n_1 + \cdots + k_p n_p = m,$$

and if m *is odd, at least one of the numbers* $n_1, \ldots, n_p$ *is also odd.*

THEOREM VI. *Given the* p *equations:*

$$f_i(y_1, \ldots, y_p, x) = 0 \qquad (i = 1, \ldots, p),$$

where the left-hand sides can be expanded in powers of x *and* y *and vanish with these variables, then, providing the equations are distinct, it is always possible to eliminate* $y_2, \ldots, y_p$ *and arrive at a unique equation* $f(y_1, x) = 0$ *of the same form as the equation in Theorem V.*

COROLLARY TO THEOREMS V AND VI. *Since Theorem IV holds whenever the Jacobian of* f *is not equal to zero, then whenever the* x *vanish,* $y_1 = \ldots = y_n$ *is a simple solution of equations* $f_1 = \ldots = f_n = 0$.

Furthermore, by Theorems V and VI and their Corollaries, Theorem IV will also still hold if the above solution is multiple, provided the order of multiplicity is odd.

In his application of the method of majorants to partial differential equations, Poincaré began with the Cauchy-Kovalevskaya theorem.[161]

In its modern form the simplest case of the theorem can be stated as follows:

Any equation of the form $\dfrac{\partial z}{\partial x} = f(x, y, z, \dfrac{\partial z}{\partial y})$ where the function f is analytic in its arguments for values near the given initial conditions $(x_0, y_0, z_0, \dfrac{\partial z}{\partial y})$ where $\dfrac{\partial z}{\partial y}$ is evaluated at $x = x_0$, $y = y_0$ possesses one and only one solution $z(x, y)$ which is analytic near (x_0, y_0).

The theorem can be generalised to functions of more than two independent variables, to derivatives of higher order and to systems of equations. This is an important result in the theory of partial differential equations which continues to play a major role today.[162]   In stating the theorem, Poincaré accorded due credit to Kovalevskaya, and the acknowledgement he gave to her here is often cited, "*Mme Kovalevskaya has considerably simplified Cauchy's proof and has given the theorem its definitive form.*"[163]

As discussed in Chapter 3 above, Poincaré himself had previously extended Kovalevskaya's results in his thesis. He now generalised these results, which concerned the first-order partial differential equation

$$\frac{\partial z}{\partial x_1} X_1 + \cdots + \frac{\partial z}{\partial x_n} X_n = \lambda_1 z,$$

where the X_i are power series in $x_1, \ldots, x_n$, to the equation

$$\frac{\partial z}{\partial t} + \frac{\partial z}{\partial x_1} X_1 + \cdots + \frac{\partial z}{\partial x_n} X_n = \lambda_1 z,$$

and found sufficient conditions for this equation to have an integral which can be expanded in powers of x and which is periodic with respect to t.

He then considered the partial differential equation

$$\frac{\partial z}{\partial t} + \frac{\partial z}{\partial x_1} X_1 + \cdots + \frac{\partial z}{\partial x_n} X_n = 0,$$

and showed that a general integral of this equation is given by

$$z = f(T_1 e^{-\lambda_1 t}, \ldots, T_n e^{-\lambda_n t}),$$

where f is an arbitrary function, and T_i are power series in x and periodic with respect to t. Furthermore since solving this partial differential equation is equivalent to solving a system of ordinary differential equations of the form

$$(14) \qquad dt = \frac{dx_1}{X_1} = \ldots = \frac{dx_n}{X_n},$$

---

[161]See Kovalevskaya [1875].

[162]A good and thorough study of Kovalevskaya's work, together with some applications of the theorem, are given by Cooke [1984, *22-38*].

[163]Poincaré [P2, *26*].

he observed that a general integral of equations (14) is given by

$$T_1 = K_1 e^{\lambda_1 t}, \ldots, T_n = K_n e^{\lambda_n t},$$

where K_i are n constants of integration.[164]

In order to determine the variables $x_1, \ldots, x_p$, as functions of $x_{p+1}, \ldots, x_n$, he considered

(15) $$\frac{\partial x_i}{\partial t} + \frac{\partial x_i}{\partial x_{p+1}} X_{p+1} + \cdots + \frac{\partial x_i}{\partial x_n} X_n = X_i \qquad (i = 1, \ldots p)$$

and showed that these equations admit a series solution in $x_{p+1}, \ldots, x_n$, and sines and cosines of multiples of t, provided the λ satisfy certain conditions. Referring to his earlier work on differential equations [1886, *172*], he was further able to show that providing the initial conditions on λ were changed in a certain way, then the equations (15) have a particular integral of the form

$$x_i = \phi_i(x_{p+1}, \ldots, x_n, t), \qquad (i = 1, \ldots, p),$$

where the ϕ can be expanded as series in $x_{p+1}, x_{p+2}, \ldots, x_n$ and sines and cosines of multiples of t.

In the case where the equations

$$dt = \frac{dx_1}{X_1} = \ldots = \frac{dx_n}{X_n}$$

are in the same form as the equations (14) except that the λ no longer satisfy the sufficiency conditions for equations (15) to have an analytic solution, then, using the previous result, Poincaré found, not a general solution, but one containing $n - p$ arbitrary constants.

5.4.2. Trigonometric series. In the final part of Chapter I of [P2], which exactly followed *Note D* in [P1], Poincaré discussed the integration of differential equations using trigonometric series. Since the topic had been the subject of several recent studies (he noted in particular the work of Floquet, Callandreau, Bruns, and Stieltjes) his treatment was not detailed but rather focused on the general case.

Using a result which he had derived in [1886c] concerning the convergence of trigonometric series, he showed that if f is an analytic function and periodic of period 2π, then it can be represented as a trigonometric series of the form

$$f(x) = A_0 + A_1 \cos x + \cdots + A_n \cos nx + \cdots + B_1 \sin x + \cdots + B_n \sin nx + \cdots,$$

which is absolutely and uniformly convergent.

He then looked for a general solution to the system of linear differential equations

(16) $$\frac{dx_i}{dt} = \phi_{i.1} x_1 + \cdots + \phi_{i.n} x_n \qquad (i = 1, \ldots, n),$$

where the n^2 coefficients $\phi_{i.k}$ are periodic functions of t of period 2π.

He began with n linearly independent solutions of equations (16),

$$x_1 = \Psi_{i.1}(t), \ldots, x_n = \Psi_{i.n}(t) \qquad (i = 1, \ldots, n),$$

---

[164]This is a standard technique for solving partial differential equations which was introduced by Lagrange and extended by Cauchy. See Kline [1972, *531-535*].

which, since equations (16) are unchanged if t is increased to $t + 2\pi$, implied the existence of the solutions

$$x_1 = \Psi_{i.1}(t + 2\pi), \ldots, x_n = \Psi_{i.n}(t + 2\pi).$$

There must be linear combinations of the original solutions, so that

$$\Psi_{i.k}(t + 2\pi) = A_{i.1}\Psi_{1.k}(t) + \cdots + A_{i.n}\Psi_{n.k}(t) \qquad (i, k = 1, \ldots, n),$$

where the A are constant coefficients. If S_1 is a root of the eigenvalue equation

$$\begin{vmatrix} A_{1.1} - S & A_{1.2} & \ldots & A_{1.n} \\ A_{2.1} & A_{2.2} - S & \ldots & A_{2.n} \\ \ldots & \ldots & \ldots & \ldots \\ A_{n.1} & A_{n.2} & \ldots & A_{n.n} - S \end{vmatrix} = 0,$$

then there are constants B_k such that

$$\theta_{1.i}(t + 2\pi) = S_1 \theta_{1.i}(t),$$

where

$$\theta_{1.i}(t) = \sum_{k=1}^{n} B_k \Psi_{k.i}.$$

If $S_1 = e^{2\alpha_1 \pi}$, then $S_1 \theta_{1.i}(t)$ will be periodic of period 2π and can be expanded as a trigonometric series $\lambda_{1.i}$. Moreover, providing the functions $\phi_{i.k}$ are analytic, then the series will be absolutely and uniformly convergent.

Hence Poincaré wrote a particular solution to the differential equations as

$$x_i = e^{\alpha_1 t}\lambda_{1.i}(t),$$

which gave a correspondence between each root of the eigenvalue equation and each particular solution of the differential equations. If all the roots of the eigenvalue equation are distinct, then there will be n linearly independent solutions to the differential equations, and Poincaré therefore expressed the general solution as

$$x_i = C_1 e^{\alpha_1 t}\lambda_{1.i}(t) + \cdots + C_n e^{\alpha_n t}\lambda_{n.i}(t),$$

where C and a are constants.

Poincaré also showed that if the eigenvalue equation has a double root, then terms of the form $e^{\alpha_1 t}t\lambda(t)$ will be introduced into the solution of the differential equations. Similarly, a triple root will introduce terms of the form $e^{\alpha_1 t}t^2\lambda(t)$, and so on.

In this analysis Poincaré was augmenting results on the theory of differential equations which had originated with Euler and Johann Bernoulli, been generalised to the complex case by Fuchs, and finally connected to the Jordan canonical form by Hamburger.[165] Poincaré's innovation was to extend the theory to a system of differential equations with periodic coefficients.

---

[165]See Gray [1984, *1-5*].

5.5. Theory of invariant integrals

Consonant with his qualitative approach to the theory of differential equations, Poincaré's investigations into the three body problem are dominated by his research into the geometry of the problem. As the first stage of this research he made a thorough analysis of the concept of *invariant integrals*, which he had originally introduced in [1886].

Although Poincaré was not the first to recognise the existence and value of invariant integrals—they are earlier encountered in both Liouville [1838] and Boltzmann [1871]—he was the first to formalise a theory centred on the concept. In [1886a] he had used the idea of a particular invariant integral within the context of a problem concerning the stability of the solutions of differential equations. He now considered the whole concept in a broader sense, developing a general theory which revealed that the existence of an invariant integral is a fundamental property of Hamiltonian systems of differential equations. Of particular importance is his use of the theory in connection with the stability of the motion in the restricted three body problem.

The last part of the chapter is devoted to a series of theorems, all of which are characterised by their geometric nature and include one of Poincaré's most celebrated results: the original formulation of his recurrence theorem. These theorems provide Poincaré with the geometrical framework for his later analysis, the qualitative study giving him an insight into the global behaviour of the system. In the introduction to *Note F* in [P1] (which does not appear in [P2]) Poincaré made a remark which gives an insight into how he himself viewed these particular theorems: *"These theorems have been given in a geometric form which has to my eyes the advantage of making clearer the origin of my ideas ..."*.[166]

The chapter is also particularly important with regard to Poincaré's error. For it was at the end of this chapter that he made the first mistake. In essence, he failed to take proper account of the exact geometric nature of a particular curve, and it was in correcting this mistake that he was forced to make dramatic changes in the geometric description of his later results.

5.5.1. Definition of invariant integrals. Poincaré considered the system of differential equations

$$(17) \qquad \frac{dx_i}{dt} = X_i,$$

where the X_i are some functions of $x_1 \ldots x_n$, and the equations are regarded as defining the motion of a point with coordinates $(x_1, \ldots, x_n)$ in an n-dimensional space. Thus, if the initial positions of an infinite number of such points form an arc of a curve C in the n-dimensional space, then at time t they will have formed a displaced arc C', its shape determined by the differential equations.

He defined an invariant integral of the system as an expression of the form $\int \sum Y_i dx_i$ which maintains a constant value at all times t, where the integration is taken over the arc of a curve and the Y_i are some functions of x. He extended

---

[166]"Ces théorèmes ont été présentés sous une forme géométrique qui avait à mes yeux l'avantage de mieux fair comprendre la genèse de mes idées ...", [P1, *220*].

the definition to encompass double and multiple integrals, where the order of the invariant integral is defined to correspond with the dimension of the region of integration.

To give a dynamical interpretation of the idea, he used the example of the motion of an incompressible fluid, where the motion of the fluid is described by the differential equations

$$\frac{dx}{dt} = X, \qquad \frac{dy}{dt} = Y, \qquad \frac{dz}{dt} = Z,$$

together with the incompressibility condition

$$\frac{\partial X}{\partial x} + \frac{\partial Y}{\partial y} + \frac{\partial Z}{\partial z} = 0.$$

Since the fluid is incompressible, the flow is volume preserving, and so the volume, which is given by the triple integral $\int \int \int dx\,dy\,dz$, is an invariant.

More generally, if the equations (17) have the added relation

$$\sum \frac{\partial X_i}{\partial x_i} = 0,$$

then the "volume" $\int \int \cdots \int dx_1 dx_2 \cdots dx_n$, is always an invariant integral. Thus the equations in Hamiltonian form

$$\frac{dx_i}{dt} = \frac{\partial F}{\partial y_i}, \qquad \frac{dy_i}{dt} = -\frac{\partial F}{\partial x_i},$$

where F is a function of the double series of variables $x_1, \dots, x_n, y_1, \dots, y_n$, and the time t, always admit the volume in phase space, $\int \dots \int dx_1 \dots dx_n dy_1 \dots dy_n$, as an invariant integral, since

$$\sum \frac{\partial}{\partial x_i} \left(\frac{\partial F}{\partial y_i} \right) + \sum \frac{\partial}{\partial y_i} \left(\frac{-\partial F}{\partial x_i} \right) = 0.$$

By considering a particular solution of the variational equations, Poincaré found a second invariant of the Hamiltonian system, namely the double integral $\int \int \sum dx_i dy_i$.

Looking specifically at the n body problem, he found that there existed not only invariant integrals which could be deduced from the ten classical integrals of the problem, but that there was also a further invariant not associated with any integral of the original equations. This additional invariant was given by

$$\int \sum (2x_i dy_i + y_i dx_i) + 3(C_1 - C_0)t,$$

where C_0 and C_1 are the values of the energy constant at the extremities of the arc along which the integral is evaluated (in $6n$ dimensional space).

The proof of these last two results first appeared in *Note C*.

5.5.2. Transformation of invariant integrals.
The transformation of variables is one of the most frequently employed methods of solving differential equations in celestial mechanics, and so it was natural for Poincaré to consider the effect of such transformations on the associated invariant integrals.

Considering the system of differential equations (17) with the condition

$$\sum \frac{\partial(MX_i)}{\partial x_i} = 0,$$

such that $J = \int M \, dx_1 \ldots dx_n$ is a positive invariant, he found that the transformation

$$x_i = \Psi_i(z_1, \ldots, z_n) \qquad (i = 1, \ldots, n)$$

left the invariant J positive, provided that in the domain under consideration the x are single-valued functions of the z and vice versa.[167]

In the case where one of the new variables is chosen to be $z_n = C$, where $F(x_1, \ldots, x_n) = C$ is a particular integral of the original equations, he found that the transformed equations

$$\frac{dz_1}{dt} = Z_1, \ldots, \frac{dz_{n-1}}{dt} = Z_{n-1},$$

admit a positive invariant integral of order $n - 1$.

He further observed that the situation was different if the transformation included the independent variable t. For if equations (17) have an invariant integral of order n, and if t_1 is the new independent variable which is defined by $t_1 = \theta(x_1, \ldots, x_n)$, then the new invariant integral is given by

$$\int M \left(\frac{\partial \theta}{\partial x_1} X_1 + \ldots + \frac{\partial \theta}{\partial x_n} X_n \right) dx_1 \ldots dx_n.$$

From this result Poincaré was led naturally to the consideration of sections transverse to the flow. For in the case where $n = 3$ and the x_i are regarded as the coordinates of a point P in space, then a transverse section S of a surface $\theta(x_1, x_2, x_3) = 0$ is the part of the surface on which all the points satisfy

$$\frac{\partial \theta}{\partial x_1} X_1 + \frac{\partial \theta}{\partial x_2} X_2 + \frac{\partial \theta}{\partial x_3} X_3 \neq 0.$$

In other words, the flow defined by the differential equations goes through the surface S and is nowhere tangent to it.

To investigate the existence of invariant integrals over S, Poincaré used his idea of consequents. In [1882] he had introduced these as point iterations on transverse sections, but he now extended the idea to include curves and areas. He considered a volume V bounded by a surface trajectory, where the surface trajectory was formed from a curve C on S bounding an area A passing to its iterate C' bounding an area A'. He showed that if there is a positive invariant integral which extends to the volume V, then there is another integral which conserves its value over the area A or any of its consequents.

---

[167]The function M satisfying the linear partial differential equation, called the *last multiplier* of the system of differential equations, was introduced by Jacobi in [1844].

5.5.3. The use of invariant integrals. To look at the role of invariant integrals in relation to the stability of the solutions of the restricted three body problem, Poincaré extended his original definition of stability to include the definition he had used in [1885] and which he now called Poisson stability. In this definition the motion of a point P is said to be stable if it returns infinitely often to positions arbitrarily close to its initial position.

Using the result that today is more familiarly known as his recurrence theorem, Poincaré established that, given certain initial conditions, there are an infinite number of solutions of the restricted problem that are Poisson stable, and that those which are not Poisson stable can be considered exceptional in a sense which he made precise.

THEOREM I (recurrence theorem). *Suppose that the coordinates x_1, x_2, x_3 of a point P in space remain finite, and that the invariant integral $\int \int \int dx_1 dx_2 dx_3$ exists; then for any region r_0 in space, however small, there will be trajectories which traverse it infinitely often. That is to say, in some future time the system will return arbitrarily close to its initial situation and will do so infinitely often.*

In other words, given a system with three degrees of freedom in which the volume is preserved, there are an infinite number of solutions which are Poisson stable.

Poincaré's proof of the theorem is attractively simple.[168]

Consider a region R with volume V within which the point P remains. Then consider a very small region r_0 of R with volume v which at time t consists of an infinite number of moving points. At time τ these points will have filled out a region r_1, at time 2τ a region r_2, etc., and at time $n\tau$ a region r_n, where r_0 and r_1 have no point in common and r_n is the nth iterate of r_0. Since the volume is preserved, each region $r_0 \ldots r_n$ will have the same volume v. Thus if $n > \dfrac{V}{v}$ then at least two of the regions have a part in common. Consideration of the successive iterates of this common region shows that there is a collection of points which belong simultaneously to r_0 and to an infinite number of other regions, and that this collection of points itself forms a region σ. From the definition of the region σ, every trajectory which starts from a point within it goes through the region r_0 infinitely often.

COROLLARY. *It follows from the above that there exist an infinite number of trajectories which pass through the region r_0 infinitely often; but there may exist others which pass through it only a finite number of times, although these latter trajectories may be regarded as exceptional.*

By exceptional Poincaré meant that the probability that a trajectory starting in the region r_0 does not pass through the region more than k times is zero, however

---

[168] It is sometimes suggested that in order properly to rigorise Poincaré's argument it is necessary to have the concept of the "measure" of a set of points, a concept which was not available until Lebesgue presented his ideas on integration in [1902]. In 1915 Van Vleck [1915, *335*] reformulated the theorem in terms of measure theory, and shortly afterwards Carathéodory [1919] provided a proof. Wintner [1947, *414*] believed Poincaré's proof to be correct, and, according to Brush [1980], this view is endorsed by Clifford Truesdell, who considers Carathéodory's reformulation to be simply "cosmetic".

large k and however small the region r_0. The corollary and its proof were additions to [P2]. In [P1] Poincaré simply stated the claim that the stable trajectories would outnumber the unstable, in direct analogy with the irrational and rational numbers.

Poincaré also pointed out that the theorem holds in a variety of other cases, namely when

i) the volume is not an invariant integral but there exists a positive invariant integral $J = \int \int \int M dx_1 dx_2 dx_3$, which remains finite;

ii) $n > 3$ providing there exists a positive n-dimensional invariant integral and the n coordinates of the point P in the n-dimensional space remain finite;

iii) the positive n-dimensional invariant integral extended over the whole n-dimensional space remains finite, even if the n coordinates are not constrained to remain finite.

He also distinguished between the cases when a known integral of equations (17)

$$F(x_1, \ldots x_n) = constant,$$

is the equation of a system of closed surfaces in an n-dimensional space, and when the integral is the equation of a system of unbounded surfaces in an n-dimensional space. In the former the conditions of the theorem are satisfied without any further constraints, but in the latter the theorem only holds providing a positive invariant integral exists which has a finite value when extended to all systems of values of x where $C_1 < F < C_2$.

In [P2] Poincaré used this last property to extend a result in Hill's lunar theory. Hill [1878] had proved the existence of an upper bound for the radius vector of the moon. Poincaré was now able to strengthen Hill's result by proving that the moon returned infinitely often to positions as close as desired to its initial position, i.e., that the moon has Poisson stability. In his proof he regarded the variables in Hill's differential equations as representing the coordinates of a point in four-dimensional space so that their accompanying integral represented a system of unbounded surfaces. He then showed that the fourth-order invariant integral of the system extended to all points contained between two of these surfaces was finite. From this it followed that his recurrence theorem held, which in turn implied the existence of trajectories passing infinitely often through any region (however small) of the four-dimensional space.

With regard to the restricted three body problem, Poincaré appealed to Bohlin's [1887] generalisation of Hill's result, in which Bohlin had proved the existence of an upper bound for the radius vector of the planetoid, to show that, providing the Jacobian integral remains within certain limits (which in general it does), the motion of the planetoid also possesses Poisson stability. He also observed that the result could not be extended to the general three body problem since it is then no longer possible to assign limits to the coordinates.

In [P1] Poincaré included very little of the above concerning Hill's theory and made no explicit statement about stability in connection with either the lunar theory or the restricted three body problem. He did not prove the result concerning the fourth-order invariant integral nor did he make its significance more accessible by putting it into the context of a particular problem. It was therefore difficult to understand what he was trying to achieve and why. He partly ameliorated the problem in *Note B*, where he translated his results into the more physical language of the astronomers and gave the relevant references to the work of Hill and Bohlin.[168a] Nevertheless, even *Note B* was not sufficiently clear on the behaviour of the radius vector to alleviate all the confusion, and Mittag-Leffler sought further clarification by asking Poincaré for a summary of his definition of stability.[169] Poincaré's detailed response, in which he carefully spelt out the differences between his results and those of Hill [1878] and Bohlin [1887] (i.e., that he had proved the existence of both lower and upper bounds for the radius vector of the planetoid), formed the basis for *An addition to Note B*, which appears at the end of *Note B* in [P1].[170]

Poincaré's next theorem is a generalisation of the result he had applied in [1886b], when he had used the idea of an invariant integral for the first time. This and the remaining results in the second chapter of the memoir are concerned with the properties of the mapping associated with the flow which takes a transverse section into itself.

THEOREM II. *If x_1, x_2, x_3, represent the coordinates of a point in space, and there exists a positive invariant integral, then there is no closed transverse section. For $n > 3$ the theorem can be given analytically.*

After the proof of Theorem II Poincaré took the step of introducing a small parameter into the differential equations, the reason being that many dynamical problems, especially those of celestial mechanics and in particular the restricted three body problem, naturally involve small parameters which are used to form power series expansions of the solutions to the differential equations.

In the restricted problem the natural parameter which arises is that of the mass of the smaller of the two primaries, generally designated by μ. The advantage of using μ as the parameter is that it is possible to change the nature of the problem by changing the value of μ. For if $\mu = 0$, the problem reduces to a pair of two body problems and can therefore be solved. This leads to the idea of starting with a particular solution for which $\mu = 0$, and then seeing if it is possible to find solutions for values of μ which are close to but not equal to zero. This is exactly what Poincaré did.

To apply his theoretical results to the restricted problem, Poincaré considered the differential equations

$$\frac{dx_1}{dt} = X_1, \qquad \frac{dx_2}{dt} = X_2, \qquad \frac{dx_3}{dt} = X_3,$$

as functions of the x_i and μ, the solutions of which can be expanded in terms of the parameter.

LEMMA I. *Consider part of a transverse section S, passing through the point a_0, b_0, c_0; if x_0, y_0, z_0 are the coordinates of a point of S, and if x_1, y_1, z_1 are the coordinates of its consequent, then x_1, y_1, z_1 can be expanded in powers of $x_0 - a_0, y_0 - b_0, z_0 - c_0$ and μ, providing these quantities are sufficiently small.*

[168a] See Chapter 5, end of Section 5.7.

[169] Mittag-Leffler to Poincaré, 21.12.1888, I M-L.

[170] Poincaré to Mittag-Leffler, 15.1.1889, No. 45a, I M-L.

Lemma I was first included in *Note E*, where its proof referred to what is now Theorem II in §5.4. In [P2] Poincaré adjusted the proof by referring to Theorem III from the same section. It made sense to insert it where it now came because of its role in the proof of the following lemma, which did appear in the main text of [P1].

LEMMA II. *If the distance between two points A_0 and B_0 belonging to part of a transverse section S is very small of nth order, then it will be the same for their iterates A_1 and B_1.*

In each version of this lemma the expression "small quantity of nth order" had a different meaning. In [P1] Poincaré defined a function of x_1, x_2, x_3 and μ as being a small quantity of nth order if it could be expanded in powers of μ with the first term in the expansion being a term in μ^n. In [P2] he defined a function of μ, which need not have a power series expansion in μ, as a small quantity of nth order if it tended to zero with μ in such a way that the ratio of the function to μ^n tended towards a finite limit. This change was needed to fit in with the subsequent alterations to Theorem III.

Up to this point Poincaré's revisions were more or less cosmetic: theorems added, proofs enhanced but no fundamental alterations to the results in [P1]. Essentially the effect has been to give this early part of the memoir a more coherent structure. The rest of this chapter of the memoir tells a rather different story. The next series of changes that Poincaré made were not simply improvements but were necessitated by the discovery of a mistake in the Corollary to Theorem III [P1]. This resulted in a significant change to the theorem itself and the removal of the Corollary altogether.

Before discussing these theorems, one definition from [P1] (which does not appear in [P2]) is needed. In Theorem III [P1] Poincaré uses a new term, *quasi-closed*, which he did not use in [P2] but which is important with regard to the error, although unfortunately his definition is not altogether clear. He said that an nth order curve C, by which he meant a curve coincident with its nth iterate, which in general is dependent on μ and is contained on part of a transverse section S, was quasi-closed if there were two points A and B on it which were separated by a finite arc but whose distance apart was very small of pth order.

THEOREM III ([P1]). *If an invariant curve C is quasi-closed such that the distance between the points of closure A and B is very small of nth order, and there exists a positive invariant integral, the distance from the point A to its iterate A_1 and that of B to its iterate B_1 are very small of nth order.*

In proving the theorem, Poincaré referred to *Figures 5.5.i* and *5.5.ii*[171] and his argument, which hinged on an application of the triangle inequality, showed that the configuration given in *Figure 5.5.ii*, as opposed to that given in *Figure 5.5.i*, was correct.

COROLLARY ([P1]). *If it has been proved that an invariant curve C is quasi-closed so that the distance between the points of closure A and B is very small of nth order at least, if moreover it is known that the distance of the point A to its*

---

[171]Poincaré [P2, *329*]

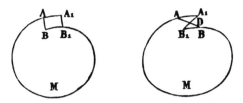

FIGURE 5.5.i   FIGURE 5.5.ii

iterate is a finite quantity or a small quantity of $n-1$th order at most, and finally if there exists a positive invariant integral, then the curve C is closed.

Poincaré gave no proof of the corollary, simply observing that if the curve was only quasi-closed, then the distance AA_1 would have to be of nth order. What he did not explore was the possibility that the curve, rather than being closed, might be self-intersecting. In essence he failed to take into account the full range of possibilities consistent with the constraint of area-preservation imposed by the existence of the invariant integral. Although he knew that the area inside the curve had to remain constant and independent of the iterative process, he focused on a single iteration and appears not to have investigated the possible outcome engendered by the extension of the iterative process. As he later realised and showed in Theorem III [P2], the concept consistent with area preservation was not closure but self-intersection.

Poincaré set up Theorem III [P2] in essentially the same way as the Corollary to Theorem III [P1].

THEOREM III ([P2]). *Let A_1AMB_1B be an invariant curve, such that A_1 and B_1 are the iterates of A and B. Suppose that the arcs AA_1 and BB_1 are very small (i.e., they tend to zero with μ) but that their curvature is finite. Suppose that the invariant curve and the position of the points A and B depend upon μ according to some rule, and that there exists a positive invariant integral. If the distance AB is very small of the nth order and the distance AA_1 is not very small of the nth order, then the arc AA_1 intersects the arc BB_1.*

He then discussed the four possible hypotheses:

1. The two arcs AA_1 and BB_1 intersect each other.

2. The curvilinear quadrilateral AA_1B_1B is such that the four arcs which comprise its sides do not have a point in common except for the four corners A, A_1, B, B_1 (as in *Figure 5.5.i*).

3. The two arcs AB and A_1B_1 intersect each other at a point D (as in *Figure 5.5.ii*).

4. One of the arcs AB or A_1B_1 intersects one of the arcs AA_1 or BB_1; but the arcs AA_1 and BB_1 do not intersect each other, neither do the arcs AB and A_1B_1.

He found that he could eliminate the second and the fourth hypotheses because they failed the condition of area preservation; in both cases the area AMB was not

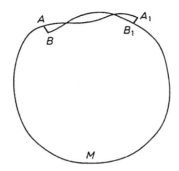

FIGURE 5.5.iii

equal to the area A_1MB_1, although in order to prove the latter case he had to do some additional juggling with the arcs. He also found that the third hypothesis was unacceptable because it implied that the distance AA_1 must be a quantity of nth order (as in Theorem III [P1]) and not a distance of $(n-1)$th order at most as specified. Thus he could conclude that the first hypothesis, that the curve was self-intersecting, was the only one possible.

Poincaré did not include a diagram, but the correct form of the curve is shown in *Figure 5.5.iii*. The reason the curve has to take this slightly complicated shape with two crossing points is to allow for more than one iteration. If after one iteration the curve had only one crossing point, then any subsequent iteration would violate the condition of area preservation.

Why did Poincaré make the mistake in [P1]? Was it simply an oversight? As previously remarked, Poincaré was renowned for paying scant attention to detail, and certainly the deadline for the competition would not have encouraged him to do otherwise. However, perhaps a more convincing argument might be that he had a preconceived idea about how he thought the curve would behave. If he thought he had found what he was expecting, he might not have felt the necessity to scrutinise his results, particularly if he felt pressed for time. As is revealed later, the behaviour of the self-intersecting curve is extremely complex and quite unlike anything Poincaré (or anybody else) had previously encountered. Indeed, when he discovered the mistake and its implication, it came as a complete shock.[172]

In [P1] Poincaré also included two extensions to Theorem III, as well as a fourth theorem. Since none of these appeared in [P2], they have been included as Appendix 6.

5.6. Theory of periodic solutions

Poincaré's discussion of periodic solutions forms the central topic of the memoir. In it he brings together principles and techniques from the previous chapters and from his earlier papers on differential equations and the three body problem. The chapter is dominated by two important ideas connected with the stability of the periodic solutions. The first of these concerns certain constants which appear

---

[172]Poincaré to Mittag-Leffler, postmarked 1.12.1889, No. 54a, I M-L. The letter is reproduced in full at the end of §5.8.3.

in the solutions and which he originally discussed in [1886]. These are now identified as *characteristic exponents*, and an investigation into their behaviour reveals information about the stability of the solutions. Secondly, there is his remarkable discovery of an entirely new class of solutions which asymptotically approach an unstable periodic solution and which he called *asymptotic solutions*.

In the rewriting Poincaré made several additions and alterations to this chapter; only the section on characteristic exponents survived the transition intact. The most radical change concerned the analytical description of the asymptotic solutions, which underwent a major revision, culminating in the addition, at the end of the chapter, of a completely new section concerned with the asymptotic solutions of the autonomous Hamiltonian equations.

5.6.1. Existence of periodic solutions. Poincaré began with the equations

$$(18) \qquad \frac{dx_i}{dt} = X_1 \qquad (i = 1, \ldots, n),$$

where the X_i are functions of x, t and the mass parameter μ, and have the additional condition that they are periodic of period 2π with respect to t. Assuming that for $\mu = 0$ there exists a periodic solution $x_i = \phi_i(t)$, where ϕ_i is a periodic function of t with period 2π, then the question Poincaré asked was whether this periodic solution could be analytically continued for small values of the disturbing parameter μ.

Poincaré began by looking for series in powers of μ with periodic coefficients that would satisfy the differential equations. If, having proved the existence of such series, he could also prove their convergence, then he would have proved the existence of the required periodic solutions. However, having got as far as proving the existence of the series, he decided that instead of proving their convergence he would prove the existence of the periodic solutions, which would then imply the convergence of the series. It is not clear why Poincaré changed his approach, especially as he said he thought that the convergence argument could be made directly, although he gave no indication as to how this could be done. Perhaps, as Ian Stewart suggests, he could foresee complications, or maybe he was not absolutely sure how to go about it.[173]

He considered a particular solution close to the original periodic solution

$$x_i(0) = \phi_i(0) + \beta_i, \qquad x_i(2\pi) = \phi_1(0) + \beta_i + \Psi_i,$$

where Ψ_i are analytic functions of μ and β which vanish with these variables, and then sought Ψ_i such that they satisfy the equations

$$(19) \qquad \qquad \Psi_1 = \ldots = \Psi_n = 0.$$

His analysis showed that, providing the Jacobian Δ of Ψ with respect to β was not zero, these equations could be resolved. In other words equations (18) have periodic solutions for small values of μ.

In considering the case when $\Delta = 0$, he used the same method in both [P1] and [P2], but the method's dependence on the method of majorants meant that his improvements to Chapter I in [P2] were particularly beneficial with respect to clarifying his procedure.

---

[173]Stewart [1989, *67*].

If equations (19) are distinct, then $\beta_1, \ldots, \beta_{n-1}$ can be eliminated to give a unique equation $\Phi = 0$. If μ and β_n are then regarded as coordinates of a point in a plane, this equation can be regarded as representing a curve passing through the origin with each of its points corresponding to a periodic solution. By constructing the part of the curve close to the origin, Poincaré was then able to study the behaviour of periodic solutions which correspond to small values of μ and β.

If $\Delta = 0$, then (for $\mu = \beta_n = 0$) $\dfrac{d\Phi}{d\beta_n} = 0$. That is, the curve $\Phi = 0$ is tangent to the line $\mu = 0$ at the origin, and, moreover, when $\mu = 0$, the equation $\Phi = 0$ will be an equation in β_n which admits zero as a multiple root. If the order of multiplicity of the root is m, then, by Theorem V of Chapter I, there exist m series in positive fractions of μ, which vanish with μ and, when substituted for β_n, satisfy $\Phi = 0$. Using these series, Poincaré considered the intersection of the part of $\Phi = 0$ which is close to the origin with the two lines $\mu = \epsilon$, $\mu = -\epsilon$ which are very close to the line $\mu = 0$. If $m_1(m_2)$ are the number of points of intersection of $\Phi = 0$ with $\mu = \epsilon$ $(\mu = -\epsilon)$ which are real and close to the origin, then Poincaré claimed that m, m_1 and m_2 all have the same parity.[174] Thus if m is odd, then m_1 and m_2 are at least equal to one, and there exist periodic solutions for small values of μ. The result holds for both positive and negative values of μ, although clearly in the context of the restricted three body problem no physical meaning can be attached to the latter.

The above analysis also led Poincaré to the important result that as μ varies the periodic solutions disappear in pairs in the same way as real roots of algebraic equations. For if $m_1 \neq m_2$, then, since they have the same parity, their difference is an even integer, and so as μ increases continuously, the number of periodic solutions which disappear as μ changes sign will be even. In other words, a periodic solution can only disappear when it becomes identical with another periodic solution.

Poincaré looked at the case when for $\mu = 0$ the differential equations admit an infinite number of periodic solutions of the form

$$x_1 = \phi_1(t, h), \ldots, x_n = \phi_n(t, h),$$

where h is an arbitrary constant. The equations (19) are no longer distinct for $\mu = 0$ and Φ contains μ as a factor, i.e., $\Phi = \mu\Phi_1$. In this case Poincaré showed that the equations still have periodic solutions for small values of μ, but only providing that when $\mu = 0$, the equation $\Phi_1 = 0$ admits $\beta_n = 0$ as a root of odd order.

In [P2] he made the additional point that in the case where the equations admit a single-valued integral $F(x_1, \ldots, x_n) = constant$, equations (19) will not be distinct unless further conditions are imposed.

Poincaré next considered the existence of periodic solutions when the functions X_i are autonomous and the periodic solutions can be of any period. In other words, if the equations have one periodic solution, they will have an infinite number. For if $x_i = \phi_i(t)$ is a periodic solution, the same will be true of $x_i = \phi_i(t + h)$, whatever the value of the constant h.

---

[174]The justification for this claim is not immediately obvious, and Poincaré later gave an explanation in [MN I, *70-71*].

If for $\mu = 0$ the equations have a periodic solution $x_i = \phi_i(t)$ of period T, and if for small values of μ

$$x_i(0) = \phi_i(0) + \beta_i, \qquad x_i(T + \tau) = \phi_i(0) + \beta_i + \Psi_i,$$

where Ψ_i are analytic functions of $\mu, \beta_1, \dots, \beta_n, \tau$, then periodic solutions will exist for small values of μ providing it is possible to resolve the n equations

$$\Psi_1 = \Psi_2 = \dots = \Psi_n = 0$$

with respect to the $n + 1$ unknowns $\beta_1, \dots, \beta_n, \tau$.

Poincaré showed that having chosen any one of the $\beta_i = 0$, then a sufficient condition for the existence of periodic solutions for small values of μ is that not all the determinants in the matrix

$$
\begin{bmatrix}
\dfrac{\partial \Psi_1}{\partial \beta_1} & \dfrac{\partial \Psi_1}{\partial \beta_2} & \cdots & \dfrac{\partial \Psi_1}{\partial \beta_n} & \dfrac{\partial \Psi_1}{\partial \tau} \\[2ex]
\dfrac{\partial \Psi_2}{\partial \beta_1} & \dfrac{\partial \Psi_2}{\partial \beta_2} & \cdots & \dfrac{\partial \Psi_2}{\partial \beta_n} & \dfrac{\partial \Psi_2}{\partial \tau} \\[2ex]
\cdots & \cdots & & \cdots & \cdots \\[2ex]
\dfrac{\partial \Psi_n}{\partial \beta_1} & \dfrac{\partial \Psi_n}{\partial \beta_2} & \cdots & \dfrac{\partial \Psi_n}{\partial \beta_n} & \dfrac{\partial \Psi_n}{\partial \tau}
\end{bmatrix}
$$

are simultaneously zero for $\mu = \beta_i = \tau = 0$, although in this case the periodic solutions have period $T + \tau$ as opposed to period T.

5.6.2. Characteristic exponents. Having established the existence of periodic solutions, Poincaré now turned his attention to the question of their stability. Assuming a periodic solution $\phi(t)$ of equations (18) had been found, he formed the variational equations in order to study the behaviour of nearby solutions.

Since the variational equations are linear differential equations with periodic coefficients, there are n particular solutions (see Chapter 1, [P2]):

$$\xi_{1.k} = e^{\alpha_k t} S_{1k}, \qquad \dots, \qquad \xi_{n.k} = e^{\alpha_k t} S_{nk}, \qquad (k = 1, \dots, n),$$

where the α are constants and the S_{ik} are periodic functions of t with the same period as $\phi(t)$.

The constants α are what Poincaré called the *characteristic exponents* of the periodic solution, and his insight was to realise that they were the key to the stability problem. For if α is purely imaginary then the ξ remain finite and the solution can be said to be stable, and, conversely, if α is not purely imaginary then the solution can be said to be unstable. In other words, investigating the stability of the periodic solutions is equivalent to investigating the properties of their characteristic exponents. As already mentioned, the idea was not entirely new to Poincaré; it had first appeared in [1886], but, as with the case of invariant integrals, he now engaged in a more detailed study.

Drawing further from his results from the first chapter in the memoir, he proceeded to show that if two characteristic exponents are equal then terms of the type $te^{\alpha_k t} S_{jk}$ appear in the solution, and, similarly, if three characteristic exponents are equal then terms which include t^2 outside the exponential and trigonometric functions appear, and so on. He also showed that if the system is autonomous or it has a single-valued integral, then in either case one of the characteristic exponents vanishes.

With regard to Hamiltonian systems, he found that the characteristic exponents can always be arranged in pairs of equal magnitude but opposite sign. Thus if the Hamiltonian system is autonomous, then two of the characteristic exponents are zero. He called the n distinct quantities α^2 the *coefficients of stability* of the periodic solution.

By considering a particular solution of the variational equations in which $\xi_i = \beta_i$ for $t = 0$, and $\xi_i = \beta_i + \Psi_i$ for $t = 2\pi$, he derived the eigenvalue equation

$$
\begin{vmatrix}
\dfrac{\partial \Psi_1}{\partial \beta_1} + 1 - e^{2\alpha\pi} & \dfrac{\partial \Psi_1}{\partial \beta_2} & \cdots & \dfrac{\partial \Psi_1}{\partial \beta_n} \\[2ex]
\dfrac{\partial \Psi_2}{\partial \beta_1} & \dfrac{\partial \Psi_2}{\partial \beta_2} + 1 - e^{2\alpha\pi} & \cdots & \dfrac{\partial \Psi_2}{\partial \beta_n} \\[2ex]
\cdots & \cdots & \cdots & \cdots \\[2ex]
\dfrac{\partial \Psi_n}{\partial \beta_1} & \dfrac{\partial \Psi_n}{\partial \beta_2} & \cdots & \dfrac{\partial \Psi_n}{\partial \beta_n} + 1 - e^{2\alpha\pi}
\end{vmatrix} = 0
$$

from which it can be seen that if $\alpha = 0$, then the equation is equivalent to $\Delta = 0$.[175] Conversely, it also implies that if Δ is zero then one of the characteristic exponents must vanish. Consequently, Poincaré could re-express his result concerning the existence of periodic solutions by saying that if equations (18) have a periodic solution for $\mu = 0$ for which none of the characteristic exponents vanish, they will have also have a periodic solution for small values of μ.

5.6.3. Periodic solutions of Hamiltonian systems.

Poincaré next considered the existence of periodic solutions in the autonomous Hamiltonian system

$$
(20) \qquad \frac{dx_i}{dt} = \frac{\partial F}{\partial y_i}, \qquad \frac{dy_i}{dt} = -\frac{\partial F}{\partial x_i}, \qquad (i = 1, 2, 3)
$$

$$
F = F_0 + \mu F_1 + \mu^2 F_2 + \dots
$$

where F_0 is a function of the x only (since in the general problem of dynamics the force function is dependent only on the distance) and $F_1, F_2, \dots$ are functions of all variables x, y and periodic of period 2π with respect to each y.

Thus when $\mu = 0$, x_i are constants and $y_i = n_i t + \varpi_i$, where $n_i = -\dfrac{\partial F_0}{\partial x_i}$, and ϖ_i are constants of integration. So for a solution of the differential equations to be periodic when $\mu = 0$, it is necessary and sufficient for the n_i to be commensurable, and, providing the $\dfrac{\partial F_0}{\partial x_i}$ are independent of each other, the x_i can always be chosen so that this condition is fulfilled. The period T will then be the lowest common

---

[175] If the solution being considered differs only slightly from the periodic solution, so that the squares and higher powers of ξ can be neglected, then the squares and higher powers of β_i can be neglected likewise.

multiple of the $\dfrac{2\pi}{n_i}$. In other words, when $\mu = 0$ there are an infinite number of choices for the constants x_i which will lead to periodic solutions.

The question then arises of whether these periodic solutions can be analytically continued for small values of μ. Poincaré found that such analytic continuation was possible providing the periodic solutions correspond (in the simplest case) to pairs of Kepler circles with rational frequency ratio and a certain phase relation determined by the critical points of a function Ψ, where Ψ is the mean value of F_1 considered as a periodic function of t.

Although he approached the question of analytic continuation using the same methods in both [P1] and [P2], the two presentations appear rather different. The results are essentially similar, but in [P2] they are expressed in a more logical order with additional explanations. In [P1] Poincaré began by expanding the coordinates as a series in μ and then proving the existence of the periodic solutions, whereas in [P2] he first proved the existence of the periodic solutions before expanding the coordinates and determining the coefficients of the series. Furthermore, [P2] contains a discussion of the application of the theory to the restricted three body problem that is not found in [P1].

Poincaré started in [P2] by supposing that for $\mu \neq 0$ a particular solution at $t = 0$ is given by

$$x_i = a_i + \delta a_i, \qquad y_i = \varpi_i + \delta \varpi_i,$$

and that for $t = T$ the solution has the values

$$x_i = a_i + \delta a_i + \Delta a_i, \qquad y_i = \varpi_i + n_i T + \delta \varpi_i + \Delta \varpi_i.$$

Thus the solution will be periodic if

$$\Delta a_1 = \Delta a_2 = \Delta a_3 = \Delta \varpi_1 = \Delta \varpi_2 = \Delta \varpi_3 = 0.$$

However, since $F = constant$ is an integral of equations (20) and F is periodic with respect to y, these equations are not independent. Hence it is only necessary to satisfy five of them. Furthermore, choosing $t = 0$ when $y_1 = 0$, gives $\varpi_1 = \delta \varpi_1 = 0$.

Poincaré showed that the five equations could be satisfied provided both that ϖ_2 and ϖ_3 were chosen in such a way that

(21) $$\frac{\partial \Psi}{\partial \varpi_2} = \frac{\partial \Psi}{\partial \varpi_3} = 0,$$

and that neither the Hessian of Ψ with respect to ϖ_2 and ϖ_3 nor the Hessian of F_0 with respect to x_i^0 were equal to zero.[176]

Since Ψ is finite and periodic in ϖ_2 and ϖ_3, equation (21) is always satisfied, and so, providing μ is sufficiently small and neither of the two Hessians vanish, there exists a periodic solution of period T, where T is determined by the choice of the numbers n_i.

---

[176]The Hessian of a function is the determinant of the matrix of which the entries are given by the second partial derivatives of the function. It is named after the German geometer Ludwig Otto Hesse (1811-1874).

Furthermore, if $n'_i = n_i(1 + \epsilon)$ then, providing ϵ is small, there exists a periodic solution for small values of μ,

$$x_i = \phi_i(t, \mu, \epsilon), \qquad y_i = \phi'_i(t, \mu, \epsilon)$$

with period $T' = \dfrac{T}{1 + \epsilon}$ which is nearly equal to T.

In the case of the restricted three body problem, where there are only two degrees of freedom, the function Ψ depends only on ϖ_2 and so the relations (21) reduce to

$$\frac{\partial \Psi}{\partial \varpi_2} = 0 \tag{22}$$

and the Hessian of Ψ reduces to $\dfrac{\partial^2 \Psi}{\partial \varpi_2^2}$. Hence, corresponding to each of the simple roots of equation (22) there is a periodic solution for all sufficiently small values of μ, and, as established earlier, the same is true for each of the roots of odd order.

Returning to the case where the periodic solutions have period T, and having shown that they could be expressed in the form of convergent series in powers of μ,

$$x_i = x_i^0 + \mu x_i^1 + \mu^2 x_i^2 + \dots \qquad (i = 1, 2, 3)$$
$$y_i = y_i^0 + \mu y_i^1 + \mu^2 y_i^2 + \dots,$$

Poincaré 's next step was to determine the coefficients of the series.

Considering the unperturbed motion gives values for x_i^0 and y_i^0, but calculating the remaining coefficients requires a careful analysis. Poincaré's procedure, although somewhat lengthy, did not, however, impose any further restrictions on the periodic solutions since the only constraint on its validity was that the Hessian of F_0 with respect to x_i^0 did not vanish.

Applying the theory to specific problems, Poincaré began with the system described by the differential equation

$$\frac{d^2 \rho}{dt^2} + n^2 \rho + m \rho^3 = \mu R(\rho, t).$$

This equation, now generally known as Duffing's equation, is often encountered in celestial mechanics, where it occurs in the theory of libration.[177] It also arises in solid mechanics, where it can be modelled by a pendulum under the action of an imposed periodic force.

To prove the existence of periodic solutions, Poincaré simply applied a series of transformations to put the equation into Hamiltonian form, and it was then straightforward to see that the requisite conditions were fulfilled. Although, as he observed, when the nonlinearity is absent, the Hessian with respect to F_0 is zero, and the theory can no longer be applied.

Turning to the three body problem Poincaré encountered the same difficulty with the vanishing Hessian, although, as he described, in the case of the restricted problem it can be easily overcome. In this particular case the small number of variables means that it is possible to find a function of F which can be used legitimately

---

[177]Duffing [1918] made an extensive study of this equation in the context of solid mechanics.

to replace F in the Hamiltonian equations and for which the Hessian of F_0 does not vanish. Unfortunately, the same method does not work in the general problem, and an alternative method of establishing the existence of periodic solutions needs to be found.[178]

5.6.4. Calculation of characteristic exponents.
To calculate the characteristic exponents of the autonomous Hamiltonian system, Poincaré began by supposing that a periodic solution of the equations was given by $x_i = \phi_i(t)$, $y_i = \Psi_i(t)$ with a nearby solution given by $x_i = \phi_i(t) + \xi_i$, $y_i = \Psi_i(t) + \eta_i$. This leads to the equations of variation

$$\frac{d\xi_i}{dt} = \sum \frac{\partial^2 F}{\partial y_i \partial x_k}\xi_k + \sum \frac{\partial^2 F}{\partial y_i \partial y_k}\eta_k, \qquad (i, k = 1, 2, 3),$$

$$\frac{d\eta_i}{dt} = -\sum \frac{\partial^2 F}{\partial x_i \partial x_k}\xi_k - \sum \frac{\partial^2 F}{\partial x_i \partial y_k}\eta_k,$$

with solutions in the form

$$(23) \qquad\qquad \xi_i = e^{\alpha t}S_i, \qquad \eta_i = e^{\alpha t}T_i,$$

S_i and T_i being periodic functions of t. Since it is an autonomous Hamiltonian system, two of the characteristic exponents are zero, and so there are only four particular solutions.

When $\mu = 0$, F is reduced to F_0, and the variational equations are reduced to

$$\frac{d\xi_i}{dt} = 0, \qquad \frac{d\eta_i}{dt} = -\sum \frac{\partial^2 F_0}{\partial x_i^0 \partial x_k^0}\xi_k,$$

where the coefficients of the second equation are constants. In this case, the most general solution is given by $\xi_i = 0$ and $\eta_i = \eta_i^0$, where η_i^0 are constants of integration, and so all six characteristic exponents are zero.

To find values for the functions α, S_i and T_i which satisfy equations (23) for small values of μ, Poincaré sought series expansions in powers of the parameter. The difficulty is that, since all the characteristic exponents are zero when $\mu = 0$, α cannot be expanded in integer powers of μ, since the conditions necessary for the implicit function theorem to be valid are no longer fulfilled. This leads to the question of whether α can be expanded in fractional powers of μ.

What Poincaré found was that α, as well as S_i and T_i, could be expanded in powers of $\sqrt{\mu}$, and so could be written[179]

$$\alpha = \alpha_1\sqrt{\mu} + \alpha_2\mu + \dots \qquad (\alpha_0 = 0, \quad \text{since } \mu = 0 \Longrightarrow \alpha = 0)$$

$$S_i = S_i^0 + S_i^1\sqrt{\mu} + S_i^2\mu + \dots$$

$$T_i = T_i^0 + T_i^1\sqrt{\mu} + T_i^2\mu + \dots.$$

To calculate the coefficients in these series, he proceeded by first substituting these series in equations (23), and differentiating with respect to t. Next, having expanded the second derivatives of F as series in integer powers of μ, he made

---

[178]Poincaré resolved this difficulty in [MN I, *133*].

[179]Poincaré was not the first to form series in powers of the square root of the parameter. As he acknowledged in his introduction to [P2], series of this type occur in Bohlin [1888], where they are used to overcome the problem of small divisors in planetary perturbation theory. Poincaré later made a careful examination of Bohlin's series in [MN II]. See Chapter 7.

the appropriate substitutions in the variational equations, and then determined the coefficients by equating powers of $\sqrt{\mu}$. By this process he was able to calculate the coefficients as far as α_i, S_i^m and T_i^m.

In [P1] Poincaré went straight into the calculation of the coefficients without first proving that such series do indeed exist, and, moreover, giving no mathematical explanation as to why they should be series in powers of $\sqrt{\mu}$ rather than μ, or indeed rather than any other fractional power of μ. He went some of the way towards rectifying this omission in *Note H*, although his proof for the existence of the series invoked theorems concerning the method of majorants which only appeared in [P2]. [P2] contained a much more detailed existence proof for the series for α, including showing that the expansion for α only contains odd powers of $\sqrt{\mu}$, and it also contained a proof for the existence of the other two series, neither of which had been included in [P1].[180]

In his determination of the coefficients in the series for α, Poincaré found that the sign of α_1^2 depended on the sign of $\dfrac{\partial^2 \Psi}{\partial \varpi^2}$; in other words, it depended on the derivative of equation (22), the roots of which correspond to periodic solutions. Since the stability of the periodic solutions depends on the sign of $|\alpha^2|$, if μ is sufficiently small, this translates into the stability being dependent on the sign of α_1^2. Poincaré was therefore interested in the behaviour of equation (22). He considered the general case when the equation only has simple roots, i.e., the roots correspond to maxima and minima of the function Ψ. Since Ψ is a periodic function, there is at least one maximum and one minimum within each period, and exactly the same number of each. Consequently there are precisely as many roots for which the derivative and α_1^2 are positive as roots for which the derivative and α_1^2 are negative. This means that, corresponding to each system of values of n_1 and n_2, there is at least one stable and one unstable periodic solution, and, providing μ is sufficiently small, there are exactly the same number of each.

In [P2] Poincaré also showed how it was possible to continue the calculation of the coefficients for the series for S_i and T_i beyond the terms S_i^m and T_i^m already calculated.

Many of the changes Poincaré made to this section can be directly attributable to intervention from Phragmén. In the introduction to [P2] Poincaré specifically mentions Phragmén's help with regard to the calculation of the coefficients for the series for S_i and T_i. In addition, according to Mittag-Leffler, Poincaré's inclusion of the existence proofs were also prompted by queries from Phragmén.[181]

5.6.5. Asymptotic solutions. Poincaré next turned his attention to the unstable periodic solutions and the behaviour of other solutions in their immediate neighbourhood.

---

[180]Although Poincaré had shown algebraically why the expansions had to be in powers of $\sqrt{\mu}$, he gave no dynamical explanation of the result. In essence, the square root arises because near a periodic solution a perturbation changes the nature of some of the phase curves, so straightforward perturbation theory cannot be used. However, a local transformation can be found in which the unperturbed Hamiltonian is similar to that of a vertical pendulum for which the separatrix (the special phase curve which separates the phase curves with different properties) width is of the order of $\sqrt{\mu}$, and this automatically introduces a $\sqrt{\mu}$ into the new perturbation.

[181]Mittag-Leffler to Poincaré 16.7.1889, I M-L.

Starting with equations (18), he supposed that

$$x_1 = x_1^0, \ldots, x_n = x_n^0$$

was a periodic solution with a neighbouring solution $x_i = x_i^0 = \xi_i$. He then derived a system of equations to determine ξ_i,

(24)
$$\frac{d\xi_i}{dt} = \Xi_i,$$

where the Ξ are functions which can be expanded in powers of ξ, are periodic with respect to t, and have no terms independent of ξ. Neglecting powers of ξ, equations (24) reduce to the linear equations of variation with general solution $\xi_i = \sum A_k e^{\alpha_k t} \phi_{ki}$, where A are constants of integration, α characteristic exponents, and ϕ periodic functions of t.

To solve the equations when they include powers of ξ, Poincaré made the linear transformation

$$\xi = \sum \eta_k \phi_{ki}$$

so that (24) become

(25)
$$\frac{d\eta_i}{dt} = H_i = H_i^1 + \ldots + H_i^n + \ldots,$$

where the H_i are functions of t and η of the same form as Ξ, and H_i^p represent the collection of terms of H_i of degree p respect to η. He then looked for general solutions to equations (24) and (25).

By writing

$$\eta_i = \eta_i^1 + \ldots + \eta_i^n + \ldots,$$

where η_i^p represent the terms of η_i of degree p with respect to A, replacing η_i in H_i^k, and calculating η_i^q by recurrence, he found

$$\frac{d\eta_i^q}{dt} - \alpha_i \eta_i^q = \sum CA^q e^{\Omega t},$$

where $A^q = A_1^{\beta_1} \ldots A_n^{\beta_n}$, $\Omega = \gamma\sqrt{-1} + \sum \alpha\beta$, γ is a positive or negative integer, $\sum \alpha\beta = \alpha_1\beta_1 + \ldots + \alpha_n\beta_n$, and the β_i are positive integers with sum q.

This equation is satisfied by

$$\eta_i^q = \sum \frac{CA^q e^{\Omega t}}{\Omega - \alpha_i},$$

where C is generally imaginary, excluding the exceptional case when $\Omega - \alpha_i = 0$, when terms in t are introduced.

He proved that the series

$$\eta_i = \sum N \frac{A_1^{\beta_1} A_2^{\beta_2} \ldots A_n^{\beta_n}}{\Pi} e^{\Omega t},$$

where Π represents the product of the divisors $\Omega - \alpha_i$, is convergent, providing that $\Omega - \alpha_i$ does not become less than any given quantity ϵ for positive integer values of β and positive or negative integers γ; i.e., if neither of the two convex polygons containing $\alpha \pm \sqrt{-1}$ contain the origin or if the real part of the quantities α are the same sign and not equal to zero. Although he observed that the convergence followed immediately from his results on the method of majorants applied to partial differential equations described earlier, he also provided a direct proof.

With the restricted three body problem in mind, Poincaré considered the particular system represented by the differential equations

$$\frac{dx_1}{dt} = X_1, \qquad \frac{dx_2}{dt} = X_2,$$

with the added condition

$$\frac{\partial X_1}{\partial x_1} = \frac{\partial X_2}{\partial x_2} = 0,$$

which implies that the "volume" is an invariant integral.

As Poincaré pointed out, since a state of the system only depends on the variables x_1, x_2, and t, it can be represented by the position of a point in space with coordinates $e^{x_1}\cos t$, $e^{x_1}\sin t$, x_2. A periodic solution can then be represented by a closed curve, and if the periodic solution is unstable the coefficient of stability α^2 will be real and positive. In this case the η_i can be expanded as a series in $Ae^{\alpha t}$ and $Be^{-\alpha t}$. Thus if $A = 0$, and $t \longrightarrow t + \infty$, then η_1 and $\eta_2 \longrightarrow 0$ and the corresponding solution asymptotically approaches the periodic solution. Similarly, if $B = 0$ and $t \longrightarrow -\infty$, then again η_1 and $\eta_2 \longrightarrow 0$ and the solution again asymptotically approaches the periodic solution. These two series of solutions, the first corresponding to $t = +\infty$ and the second corresponding to $t = -\infty$, are what Poincaré called *asymptotic solutions*. Moreover, since each of these series corresponds to a sequence of curves which asymptotically approaches a closed curve C, Poincaré called the surface formed by the set of these curves an *asymptotic surface*. Thus, there are two asymptotic surfaces, one corresponding to $t = +\infty$ and the other corresponding to $t = -\infty$, and both of these surfaces pass through the closed curve C.

In [P1] Poincaré went through a similar analysis to show that in the case of equations (18) the series for η could be expanded in a convergent series in $A_i e^{\alpha_i t}$. But at the end of the analysis he added the claim that if the differential equations depend on the parameter μ then the series could also be expanded in powers of μ or $\sqrt{\mu}$, according to the circumstances. Nowhere did he prove that such expansions were actually possible. Furthermore, implicit in his claim was that the series in each case was convergent. Particularly significant is the fact that he made no attempt to distinguish between the autonomous and nonautonomous cases. As he later discovered, neglecting to make this distinction was a serious oversight.

In [P2] the ending of the section was quite different. Poincaré proved that the η could be represented by a series in μ, and, moreover, that these series were convergent, subject to three conditions. First, that the differential equations are dependent on the parameter μ and the functions X_i can be expanded in powers of the parameter; second, that for $\mu = 0$, all the characteristic exponents α are distinct and can be expanded in integer powers of μ; and third, it is possible to remove all the constants A which correspond to an α whose real part ≤ 0.

The significant condition to observe is the second one, which concerns the characteristic exponents. For this condition means that if the system under consideration is an autonomous Hamiltonian system, then the series are not convergent, and in particular the series are not convergent in the case of the restricted three body problem. This point is central with regard to Poincaré's error. For in [P1] Poincaré had not appreciated that, in describing the behaviour of asymptotic solutions, there was a critical difference between autonomous and nonautonomous

systems, a difference which initially manifests itself in the values of the characteristic exponents. In [P2] the distinction between the two cases is clearly made, the
nonautonomous case having been dealt with here and the autonomous case being
the subject of the next section.

5.6.6. Asymptotic solutions of Hamiltonian systems.
Poincaré had al-
ready proved that there were circumstances under which the autonomous Hamiltonian system had periodic solutions, and so to establish the existence of asymptotic
solutions he only had to make certain that one of the corresponding characteristic
exponents α was real. That being so, it only remained to ascertain the form of the
asymptotic solutions.

In the case discussed in the previous section the functions X_i were expanded in
powers of μ, and the characteristic exponents were distinct for $\mu = 0$. In the case
of the autonomous Hamiltonian equations, the right-hand side of the equations can
again be expanded as powers of μ, but now all the characteristic exponents vanish
when $\mu = 0$.

This results in several important differences. First, as Poincaré had already
described, the expansions for the characteristic exponents are in powers of $\sqrt{\mu}$ rather
than μ. Similarly, the expansions of the functions $\phi_{i,k}$ which appear in the general
solution to the variational equations and which, in this case, are the expansions of
the functions S_i and T_i are also in powers of $\sqrt{\mu}$ rather than μ. Furthermore, this
implies that the expansions of the functions H_i are in powers of η, $e^{t\sqrt{-1}}$, $e^{-t\sqrt{-1}}$,
and $\sqrt{\mu}$ (and not of μ). Although η_i can be derived as before,

$$\eta_i = \sum N \frac{A_1^{\beta_1} A_2^{\beta_2} \ldots A_n^{\beta_n}}{\Pi} e^{\Omega t},$$

the expansions of N and Π are now also in powers of $\sqrt{\mu}$.

These differences led Poincaré to ask the following questions:

1. Since N and Π can be expanded as powers of $\sqrt{\mu}$, can the quotient $\dfrac{N}{\Pi}$
also be expanded in powers of $\sqrt{\mu}$?

2. If the answer to Question 1 is yes, then this implies the existence of series
in $\sqrt{\mu}$, $A_i e^{\alpha_i t}$, $e^{t\sqrt{-1}}$, and $e^{-t\sqrt{-1}}$, which *formally* satisfy the equations; are
these series convergent?

3. If the series are not convergent, can they be used to approximate the
asymptotic solutions?

With regard to Question 1, since both N and Π can be expanded in powers of
$\sqrt{\mu}$, Poincaré realised that the only problem that could arise with the expansion of
their quotient is the appearance of negative powers of $\sqrt{\mu}$. For if this should occur
then the asymptotic solutions would cease to exist for $\mu = 0$. His answer therefore
consisted in proving that these negative powers never arise. He had previously
recognised the existence of this particular problem and included an earlier version
of the proof in *Note I*.

Poincaré had therefore proved the existence of series which formally satisfied
the equations, but were these series convergent? Importantly, Poincaré showed that

they were not. However, the discovery of their divergence was entirely unexpected, since his analysis in [P1] had led him to believe that they were convergent. Put in the context of the whole memoir, his failure to appreciate the divergence of these series is essentially the analytical analogue of the geometrical mistake that he made at the end of his discussion on invariant integrals in [P1].

In [P2] Poincaré proved that, rather than being convergent, the series belonged to the class of divergent series that he had defined in [1886a] as asymptotic series. [P1a] reveals that he was slightly concerned about the status of this particular proof despite describing it in [P2] as "rigoureuse". In [P1a] the word "rigoureuse" was originally preceded by the word "plus", which was then crossed out and replaced by the word "absolument", which was also crossed out.

He began with the expression $(\Omega - \alpha_i)^{-1}$. If γ is not equal to zero, then this expression can be expanded in powers of $\sqrt{\mu}$, but the radius of convergence of the series will tend to zero as $\dfrac{\gamma}{\sum \beta}$ tends to zero. Thus if the expression $\dfrac{1}{\Pi}$ is expanded in powers of $\sqrt{\mu}$, there will always be an infinite number of such expressions for which the radius of convergence of the expansion is arbitrarily small. If the same is true for $\dfrac{N}{\Pi}$, then this implies that the series are divergent.

Rather than considering the series for η_i, Poincaré began with the simpler series

$$F(w, \mu) = \sum \frac{w^n}{1 + n\mu},$$

where $w = A_i e^{a_i t}$. This series in w is uniformly convergent when $\mu > 0$ and $|w| < w_0 < 1$, and if differentiated the resulting series is also uniformly convergent.

On the other hand, if the function $F(w, \mu)$ is expanded as a series in μ,

$$(26) \qquad F(w, \mu) = \sum_{n,p} w^n (-n)^p \mu^p,$$

then, as Poincaré knew from his theory developed in [1886a], the series is not convergent but is an asymptotic expansion.

Poincaré claimed that the series (26) was completely analogous both to the series which represent the functions η_i,

$$\sum \frac{N}{\Pi} w_1^{\beta_1} \dots w_k^{\beta_k} e^{\gamma t \sqrt{-1}} = F(\sqrt{\mu}, w_1, \dots, w_k, t), \qquad (w_i = A_i e^{\alpha_i t}),$$

and to the series

$$\sum w_1^{\beta_1} \dots w_k^{\beta_k} e^{\gamma t \sqrt{-1}} \frac{d^p \left(\frac{N}{\Pi} \right)}{(d\sqrt{\mu})^p} = \frac{d^p F}{(d\sqrt{\mu})^p}.$$

These two series are uniformly convergent when expanded in powers of w provided $|w| < w_0 < 1$ and $\sqrt{\mu}$ is real, but if $\dfrac{N}{\Pi}$ is expanded in powers of $\sqrt{\mu}$, then they are divergent. Thus if they are analogous to the series (26) they must be asymptotic expansions.

Poincaré first defined $\Phi_p(\sqrt{\mu}, w_1, \dots, w_k, t)$ to be a polynomial of degree p in $\sqrt{\mu}$ which can be expanded in powers of w, and $e^{\pm t \sqrt{-1}}$. The series for $\dfrac{F - \Phi_p}{\sqrt{\mu^p}}$ is then given by

$$\sum \frac{1}{\sqrt{\mu^p}} \left(\frac{N}{\Pi} - H_p \right) w_1^{\beta_1} \dots w_k^{\beta_k} e^{\gamma t \sqrt{-1}},$$

where H_p is the group of terms in the expansion of $\dfrac{N}{\Pi}$ in which the exponent of $\sqrt{\mu}$ is at most equal to p. To prove that the series for η_i is an asymptotic expansion, this series must be shown to be uniformly convergent with its terms tending to zero as μ tends to zero. This convergence proof turned out to require a long and delicate analysis, and Poincaré's attempt in [P2] included some unproven assertions, which doubtless accounts for his concern about the rigour.[182]

Nevertheless, he correctly concluded that the series for the asymptotic solutions

$$x_i = x_i^0 + \sqrt{\mu x_i^1} + \mu x_i^2 + \dots, \qquad y_i = n_i t + y_i^0 + \sqrt{\mu y_i^1} + \mu y_i^2 + \dots$$

were asymptotic expansions and, in addition, that if they were differentiated they would also give asymptotic expansions.

5.7. Study of the case with two degrees of freedom

In [P1] the opening chapter of the second part of the memoir consisted of five sections and constituted the major part of this half of the memoir. In [P2] Poincaré changed the structure so that only the first of these sections, which concerned the geometric representation of systems of differential equations, was contained in the opening chapter of the second part.

Poincaré's task now was to apply the foregoing theory to the restricted three body problem. He therefore focused on the Hamiltonian system with two degrees of freedom

$$(27) \qquad \frac{dx_i}{dt} = \frac{\partial F}{\partial y_i}, \qquad \frac{dy_i}{dt} = -\frac{\partial F}{\partial x_i}, \qquad (i = 1, 2),$$

where the x_i are linear variables, y_i are 2π periodic angular variables, and F is an autonomous function of x_i and y_i.

His strategy was to begin by showing how such a system can be given a geometric representation that uniquely identifies its each and every state, the chosen representation dependent on the given constraints of the particular problem under consideration.

First, Poincaré used the property that the four variables are linked by the Jacobian integral

$$F = (x_1, x_2, y_1, y_2) = C,$$

to create a representation in which each state of the system is represented by a point in a three-dimensional space. By adding further conditions he developed representations in which each state of the system was represented either by a point contained between two tori or by an interior point of a torus.

In the case of the restricted three body problem he started by defining the position of the planetoid using the osculating elements, i.e., the variables defined

---

[182]Later, in the first volume of the *Méthodes Nouvelles*, Poincaré gave a fuller version of the proof in which he supplied the missing details [MN I, *353-382*].

by means of the instantaneous ellipse described by the planetoid round the centre of gravity of the system. He adopted Tisserand's [1887] notation (derived from Delaunay) to write the equations of motion in canonical form

(28)
$$\frac{dL}{dt} = \frac{\partial R}{\partial l}, \qquad \frac{dl}{dt} = -\frac{\partial R}{\partial L}$$
$$\frac{dG}{dt} = \frac{\partial R}{\partial g}, \qquad \frac{dg}{dt} = -\frac{\partial R}{\partial G},$$

where l is the mean anomaly of the planetoid, g is the longitude of its perihelion, n is the mean motion of the planetoid, $L = \sqrt{a}$, where a is the semi-major axis of the instantaneous ellipse, and $G = \sqrt{a(1 - e^2)}$, where e is the eccentricity of the ellipse. In order to preserve the canonical form, the standard perturbation function is increased by the addition of the term $\frac{1}{2a} = \frac{1}{2L^2}$ to give the Hamiltonian R.

He chose his units so that the masses of the two primaries were $1 - \mu$ and μ, the gravitational constant was equal to one, the mean motion of the smaller of the two primaries was equal to one, and its longitude was equal to t. Under these conditions the angle from which the distance between the two smaller masses is seen from the larger differs from $l + g - t$ by a periodic function of l of period 2π.

Since the distance between the primaries is constant and the distance between the larger of the two primaries and the planetoid is only dependent on L, G and l, the function R is only dependent on L, G, l and $l + g - t$. Moreover, since R is periodic with period 2π with respect to l, and with respect to $l + g - t$,

$$\frac{\partial R}{\partial t} + \frac{\partial R}{\partial g} = 0,$$

and equations (28) admit the integral $R + G = constant$.

However, R has an explicit dependence on t, and so equations (28) are not in the required form of equations (27). To remedy this Poincaré made the transformation

$$x_1 = G, \qquad x_2 = L, \qquad y_1 = g - t, \qquad y_2 = l$$
$$F(x_1, x_2, y_1, y_2) = R + G.$$

The function F is dependent on the mass parameter μ, and so can be written

$$F = F_0 + \mu F_1,$$

which if $\mu = 0$ reduces to

$$F = F_0 = \frac{1}{2a} + G = x_1 + \frac{1}{2x_2^2},$$

which is a function of only the linear variables.

By definition, $L^2 \geq G^2$, which implies that $x_2 \geq x_1 \geq -x_2$. If $x_1 = +x_2$, then the eccentricity is zero, and the perturbation function and the state of the system only depend on the difference of the longitude of the two smaller masses, i.e., they only depend on

$$l + g - t = y_1 + y_2.$$

Consequently

$$\frac{\partial F}{\partial y_1} = \frac{\partial F}{\partial y_2},$$

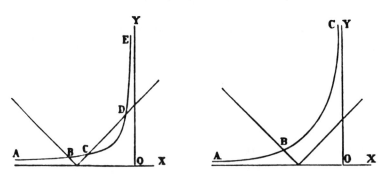

FIGURE 5.7.i   FIGURE 5.7.ii

which implies

$$\frac{d(x_1 - x_2)}{dt} = 0,$$

and since $x_2^2 \geq x_1^2$, the maximum value of $x_1 - x_2$ is 0. (x_1 is not identically equal to x_2 since in equations (27) $x_1 = x_2$ only if there is a singularity.) If $x_2 = -x_1$, again the eccentricity is zero but the motion is then retrograde, which always occurs when x_1 and x_2 are of different sign.

To create a geometric representation for the restricted three body problem, Poincaré had to represent the system using only three variables. He therefore sought to express x_1 and x_2 as single-valued functions of y_1, y_2 and a new variable ξ.

He began with $\mu = 0$ and considered the plane in which the coordinates of a point are defined by

$$X = x_1 - C, \qquad Y = x_2,$$

which, from the definition of F and the constraint on x_1, implies

$$X + \frac{1}{2Y^2} = 0, \qquad Y > X + C > -Y.$$

The construction of the curve $X + \dfrac{1}{2Y^2} = 0$, together with the lines $X + C = \pm Y$, takes two different forms depending on the value of the constant C as shown in *Figures 5.7.i* and *5.7.ii*,[183] the transition point that occurs when the line CD becomes tangent to the curve takes place when $X = \frac{1}{2}$, $Y = 1$, and $C = \frac{3}{2}$. Although the inequalities are satisfied by the curves BC and DE, in *Figure 5.7.i* and BC in *Figure 5.7.ii*, the part of the curve which is of interest with respect to the problem is the part which is bounded, that is, BC in *Figure 5.7.i*.

For $\mu = 0$, choosing $\xi = \dfrac{x_2 - x_1}{x_2 + x_1}$ fulfils the required conditions since along the arc CB, ξ increases constantly from 0 to ∞.

For small values of μ, ξ can be chosen in the same way, but only if $x_2 > 0$, and the Jacobian $\dfrac{\partial(\xi, F)}{\partial(x_1, x_2)}$ is not equal to zero. Providing the value of C is not close

---

[183]Since the figures are symmetric with respect to the X axis, Poincaré only included diagrams of the top half of the X, Y plane [P2, *402*].

to $\frac{3}{2}$ these conditions are satisfied for small values of μ, and ξ can be taken as the independent variable.

Finally, in order to make a more convenient representation, Poincaré made a further transformation to another set of canonical variables

$$x_1' = x_1 + x_2, \qquad x_2' = x_1 - x_2,$$

$$y_1' = \frac{1}{2}(y_1 + y_2), \qquad y_2' = \frac{1}{2}(y_1 - y_2).$$

In this form y_i' are angular variables which if increased by 2π generate an identical increase in y_i, and so the system remains unchanged. The system also remains unchanged if simultaneously y_1' and y_2' are each increased by π. A state of the system can then be represented by a point in space with rectangular coordinates

$$X = \cos y_1' e^{\xi \cos y_2'} \qquad Y = \sin y_1' e^{\xi \cos y_2'} \qquad Z = \xi \sin y_2'.$$

In this representation each point in space corresponds to a single state of the system, while the two systems of values (x_1', x_2', y_1', y_2') and $(x_1', x_2', y_1' + \pi, y_2' + \pi)$, which correspond to two different points of space, correspond to only one state of the system.

In addition, applying the transformation has the effect of reducing the fourth-order invariant integral of the Hamiltonian equations to a third-order positive invariant.

When $\mu = 0$, equations (28) integrate to give

$$L = constant, \qquad G = constant, \qquad g = constant, \qquad l = nt + constant.$$

These solutions can be represented by trajectories that are closed whenever the mean motion n is a rational number. They lie on the surface trajectories that are defined by the general equation $\xi = constant$ and, consequently, generate closed surfaces of revolution analogous to tori.

In the following chapter in [P2], Poincaré showed the effects on these results when the system no longer remains unperturbed and μ takes on small values.

He concluded the current chapter with the consideration of two more dynamical problems. For the first he returned to the system described by Duffing's equation, and for the second he considered a heavy point mass moving on a frictionless surface in the neighbourhood of a stable equilibrium. In each case he generated a similar representation to the one he had derived for the restricted problem.

By putting this section on geometrical representations into a chapter on its own in [P2], Poincaré has accorded it a higher degree of prominence than its counterpart in [P1] (which was included in a chapter with other sections). This change of emphasis is quite revealing.

It is clear that for Poincaré framing dynamical problems geometrically came naturally (as exemplified by his remarks on the theorems concerning invariant integrals). However, his kind of geometric approach to celestial mechanics represented something quite new in mathematics, and its sheer novelty would probably have been sufficient to make the memoir almost inaccessible to those of a more practical persuasion. This was certainly the view adopted by Mittag-Leffler, who, while studying the original memoir, expressed the concern that Poincaré's resolution of the restricted three body problem was given in a form that would be difficult to understand by anyone except those very familiar with his work. In particular he thought that astronomers would not understand it all.[184] Mittag-Leffler identified as the main source of potential difficulty the fact that Poincaré was working in a three-dimensional *multiplicité* (= manifold) which was not the Euclidean three-dimensional space in which the bodies actually moved.

Poincaré responded to Mittag-Leffler's remarks by translating his most important results into a more traditional format known to be familiar to astronomers, which he added as *Note B*. However, the discovery of the mistake invalidated a large part of the *Note*, and he completely excised all trace of it in the revision.

It could be argued therefore that he chose to use the structure of the memoir to stress the geometrical representation, as opposed to having to revise *Note B*. Given the nature of the new results, the rewriting would have been a delicate undertaking and not one he would have relished in the time he had available. In any event, since he would have been primarily concerned with the response from mathematicians rather than astronomers, it is perhaps not surprising that he chose not to develop this particular side of the problem any further at this stage. However, he did not entirely forget the astronomers, for in the following year he wrote a summary of his results from [P2]:

> ... *for the readers of the Bulletin astronomique who do not have time to read 'in extenso' the original memoir which is very voluminous.*[185]

This summary, although materially similar to *Note B*, was in fact a new paper which had a quite different structure.

5.8. Study of asymptotic surfaces

In Poincaré's geometric representation of the restricted problem, a generating unstable periodic solution and its accompanying family of asymptotic solutions are represented in the three-dimensional solution space by a closed curve and two asymptotic surfaces. In order to understand the behaviour of these asymptotic solutions, he sought the exact equations for these asymptotic surfaces. Although he approached the problem in a similar way in both versions, in [P2] he added an entirely new section simply for the purpose of stating the problem and outlining a strategy for dealing with it—a clear indication of the importance he now attached to the topic.

He first noted that it was possible to move from one surface to the other by changing the sign of the parameter μ in the equations for the surfaces. So by making such a sign change it is possible to generate the second surface from the first. Furthermore, since these two surfaces cut another, they can be considered together as two sides of the same surface. This surface will then have the special

---

[184]Mittag-Leffler to Poincaré 15.11.88, I M-L.
[185]Poincaré [1891, *480*]

feature of a double curve which identifies the particular series which satisfy the equations for the asymptotic surfaces.

The equations of these surfaces are of the form

$$\frac{x_2}{x_1} = f(y_1, y_2),$$

where x_1 and x_2 are given by the asymptotic series

(29) $x_1 = s_1(y_1, y_2, \sqrt{\mu})$ $x_2 = s_2(y_1, y_2, \sqrt{\mu})$

and satisfy

(30) $$\frac{\partial F}{\partial x_1} \frac{\partial x_i}{\partial y_1} + \frac{\partial F}{\partial x_2} \frac{\partial x_i}{\partial y_2} + \frac{\partial F}{\partial y_i} = 0.$$

In order to calculate the equations of the surfaces exactly, Poincaré proceeded in three stages. In the first approximation he calculated the first two coefficients, which, since the series were in powers of $\sqrt{\mu}$, gave an approximation with an error of the order of μ. In the next stage he considered a larger but finite number of coefficients, which give an error of the order μ^p for any fixed p, no matter how large. In the final stage he calculated the exact equations.

5.8.1. First approximation. Most of the section entitled *First approximation* in [P2] came from two consecutive sections in Chapter I [P1]. It began with *The Equation of Asymptotic Surfaces* and ended with the first half of *The Construction of Asymptotic Surfaces (first approximation)*.

Since Poincaré was concerned with the restricted three body problem, he began with the Hamiltonian equations (27), assuming that F could be expanded in powers of the mass parameter μ,

$$F = F_0 + \mu F_1 + \mu^2 F_2 + \dots,$$

with F_0 independent of y. In order to ensure the existence of a generating periodic solution, he supposed that for certain values of x_i, say x_i^0, $\dfrac{\partial F_0}{\partial x_i}(= n_i)$ were commensurable.

This gives

$$x_i = \Phi_i(y_1, y_2), \qquad (i = 1, 2)$$

as the general form of the equation of a surface trajectory, providing the functions Φ_i are chosen such that $F(\Phi_1, \Phi_2, y_1, y_2) = C$, and that they satisfy equations (30).

To integrate equations (30) Poincaré supposed

(31) $$x_i = x_i^0 + \sqrt{\mu}\, x_i^1 + \mu x_i^2 + \dots,$$

where x_i are very close to x_i^0, the latter having been chosen such that the ratio $n_1 : n_2$ is commensurable. It then remained to determine the coefficients x_i^k such that when the series (31) are substituted into the equations (30) the equations are formally satisfied.

To generate a sequence of equations from which he could determine x_i^k, Poincaré substituted the series for x_i in the series for F, then equated powers of μ.

In his determination of the coefficients x_i^k Poincaré first showed that they were periodic functions of y_1 and as such could be expanded as trigonometric series in sines and cosines of multiples of y_1. He then found that

$$(32) \qquad x_1^1 = 0, \qquad x_2^1 = \sqrt{\frac{2}{N}([F_1] + C_1)},$$

where the notation $[F_1]$ represents the average value of the function F_1 considered as a periodic function of y_1, $N = -\dfrac{\partial^2 F_0}{(\partial x_2^0)^2}$ and C_1 is an integration constant. Thus he could write the series to be used in the first approximation as

$$x_1 = x_1^0, \qquad x_2 = x_2^0 + \sqrt{\frac{2\mu}{N}([F_1] + C_1)}.$$

At this point in [P2] Poincaré stopped following the section on the *Equation of Asymptotic Surfaces* from [P1] and continued instead with material taken from the section on *Construction of Asymptotic Surfaces (first approximation)*.

The remaining pages from the *Equation of Asymptotic Surfaces*, which did not appear in [P2], contained an outline of Poincaré's method for determining the remaining coefficients x_i^k, which was not required for the first approximation and which, for each coefficient, involved the choice of an arbitrary constant of integration C_k. The section concluded with three points raised for discussion:

1. When are the series thus obtained convergent?
2. How should the arbitrary constants $C_1, C_2, \ldots C_{k-1}, \ldots$ be determined?
3. What are the properties of the functions defined by the series?

He addressed all of these points in later sections in his study of asymptotic surfaces.

In [P2] Poincaré considered his results in the context of the geometric representation of the restricted problem. To simplify the notation, he suppressed the primes and called the variables x_i and y_i (not to be confused with the original x_i and y_i, i.e., G, L, $g - t$, and l). The new y_i are linear functions of

$$y_1' = \frac{1}{2}(g - t + l) \qquad y_2' = \frac{1}{2}(g - t - l),$$

and the ratio $\dfrac{x_2}{x_1}$ is a linear function of ξ. Poincaré was now able to define completely the position of a point P in the space so that every relation between y_1, y_2, and the ratio $\dfrac{x_2}{x_1}$ was the equation of a surface, and both y_1 and y_2 could be increased by a multiple of 2π without changing the position of P.

The coefficients from (32) then gave the first approximation for the equation of the surface trajectories

$$(33) \qquad \frac{x_2}{x_1} = \frac{x_2^0 + x_2^1\sqrt{\mu}}{x_1^0 + x_1^1\sqrt{\mu}} = \frac{x_2^0}{x_1^0} + \frac{\sqrt{\mu}}{x_1^0}\sqrt{\frac{2}{N}([F_1] + C_1)}.$$

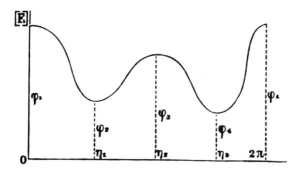

FIGURE 5.8.i

Poincaré's next problem was to identify the particular surfaces which, as he had earlier described, displayed a double curve. This led him to consider the intersections of the surfaces defined by equations (33) with the transverse section S defined by the surface $y_1 = 0$.

The position of a point P on the surface S is defined by the two coordinates $\dfrac{x_2}{x_1}$ and y_2, which, since they are analogous to polar coordinates, means that the curves $\dfrac{x_2}{x_1} = constant$ are closed concentric curves on the surface S and the position of a point P on S is unchanged when y_2 is increased by 2π. Since $\sqrt{\mu}$ is very small, the intersections of the surfaces defined by equation (33) with the transverse section defined by $y_1 = 0$ differ very little from the curves $\dfrac{x_2}{x_1} = constant$.

In order to investigate the curves formed by the intersections, Poincaré needed to understand the nature of the function $[F_1]$. He found that it was a finite periodic function of y_2. In other words, it was similar to the function Ψ he had found previously. To look at a general function of this type he supposed that as y_2 varied from 0 to 2π, $[F_1]$ varied as in *Figure 5.8.i*,[186] where $\phi_1 > \phi_3 > \phi_2 > \phi_4$.

He constructed a set of curves defined by

$$y_1 = 0, \qquad \frac{x_2}{x_1} = \frac{x_2^0}{x_1^0} + \frac{\sqrt{\mu}}{x_1^0}\sqrt{\frac{2}{N}([F_1] + C_1)},$$

the shape of each curve depending on the value of the constant C_1 as shown in *Figure 5.8.ii*.[187] Each one of these curves lies in the plane $y_1 = 0$, and so if y_1 is now varied from 0 through to 2π, the curves will each sweep out a surface. More precisely, if through each point on an arbitrary one of these curves is drawn one of the lines defined by the equations $y_2 = constant$, $\dfrac{x_2}{x_1} = constant$, then the set of all these lines constitutes a closed surface which is exactly one of the surfaces defined by equation (33).

Since each of the roots of the equation $\dfrac{[dF_1]}{dy_2} = 0$ corresponds to a periodic solution *(cf.* equation (22)), the periodic solutions correspond to the extremum

---

[186]Poincaré [P2, *418*].
[187]Poincaré [P2, *419*].

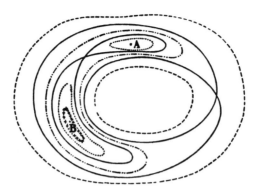

FIGURE 5.8.ii

points of $[F_1]$. In this case (due to the choice of $[F_1]$) there are four extremum points, which represent four periodic solutions, two stable and two unstable. The two stable periodic solutions correspond to the two isolated closed curves of the surfaces $C_1 = -\phi_3$ and $C_1 = -\phi_1$ (points A and B), and the two unstable periodic solutions correspond to the double curves of the surfaces $C_1 = -\phi_2$ and $C_1 = -\phi_4$. By the criteria established earlier, the latter two are the ones in which Poincaré was interested and which represent his first approximation.

In [P1], Poincaré arrived at the same result, but, due to his earlier analysis and his (erroneous) belief that the asymptotic surfaces could be represented by convergent series in $\sqrt{\mu}$, he also included the following conclusions:

1. At the first approximation the asymptotic surfaces are closed surfaces, and this result is confirmed by following approximations.

2. Since every asymptotic surface is a surface trajectory, its intersection with the transverse section S will be an invariant curve C. Consider a curve C':

$$y_1 = 0, \qquad \frac{x_2}{x_1} = \frac{x_2^0}{x_1^0} + \frac{\sqrt{\mu}}{x_1^0}\sqrt{\frac{2}{N}([F_1] - \phi_4)}.$$

This will differ very little from the invariant curve C (up to the order of μ). Its iterate will also differ very little from the iterate of C, i.e., C itself. Thus the curve C' will differ very little from its own iterate (up to the order of μ).

3. The curve C' is a closed curve; the curve C from which it differs only slightly will thus be a quasi-closed curve such that the distance between the points of closure will be of the order of μ. Thus the asymptotic surface cuts the surface $y_1 = 0$ as a quasi-closed curve.

The distance of an arbitrary point P on the surface $C_1 = -\Phi_4$ to its iterate P' will be of order $\sqrt{\mu}$. Likewise the distance of an arbitrary point on the curve C' to its iterate will also be of order $\sqrt{\mu}$.

Later it is shown how Poincaré used these results in [P1], and how they became invalidated by the discovery of the error.

5.8.2. Second approximation. The purpose of the second approximation was to determine some arbitrary number of coefficients of the series (31). Since Poincaré had originally believed the series to be convergent rather than asymptotic, there was no equivalent section in [P1].

However, most of the section is in fact taken from *Note F*, which Poincaré had added to [P1] because he wanted to include an analytic description of the asymptotic surfaces as a complement to his geometric one. *Note F*, therefore, contained what Poincaré then believed to be a description of the entire series.

Poincaré began the second approximation in the same fashion as the first, but he then transformed the problem using Hamilton-Jacobi theory.[188]

Since the system of differential equations is an autonomous Hamiltonian system, the expression $x_1 dy_1 + x_2 dy_2$ is an exact differential and so can be written

$$dS = x_1 dy_1 + x_2 dy_2,$$

where $S(y_1, y_2)$ is a solution of the Hamilton-Jacobi partial differential equation

$$F\left(\frac{\partial S}{\partial y_1}, \frac{\partial S}{\partial y_2}, y_1, y_2\right) = C.$$

S can then be expanded as a series in $\sqrt{\mu}$,

$$S = S_0 + S_1\sqrt{\mu} + S_2\mu + S_3\mu\sqrt{\mu} + \dots,$$

with coefficients S_i functions of y_1 and y_2. Moreover, since

$$\frac{\partial S_k}{\partial y_1} = x_1^k, \qquad \frac{\partial S_k}{\partial y_2} = x_2^k,$$

the problem of determining the coefficients of the asymptotic series amounts to determining partial derivatives of the coefficients in the series for S, and hence determining the coefficients in the series for S.

When $C_1 > -\phi_4$ Poincaré proved that $\dfrac{\partial S}{\partial y_1}$ and $\dfrac{\partial S}{\partial y_2}$ could be determined as (divergent) trigonometric series in sines and cosines of multiples of y_1 and y_2. But his main concern was with the case $C_1 = -\phi_4$ when the series represent the asymptotic solutions. In this case, the expression $[F_1] + C_1$ is never negative, and it only reaches zero when $y_2 = \eta_3$.

Choosing η_3 as the origin for y_2, he put the expression into the form of a trigonometric series

$$[F_1] + C_1 = \sum A_m \cos my_2 + \sum B_m \sin my_2.$$

For $y_2 = 0$ this function and its derivative vanish. Since the function is always positive, zero is, therefore, a minimum. As a result, the function

$$\frac{[F_1] + C_1}{sin^2 \dfrac{y_2}{2}}$$

---

[188]See Jacobi [1866].

can also be expanded in sines and cosines of multiples of y_2, and since it is a periodic function of y_2 which neither vanishes nor becomes infinite, it is possible to write

$$\frac{\sin \dfrac{y_2}{2}}{\sqrt{[F_1] + C_1}} = \sum A_m \cos m y_2 + \sum B_m \sin m y_2.$$

From this expression Poincaré showed that $\dfrac{\partial S_p}{\partial y_1}$ and $\dfrac{\partial S_p}{\partial y_2}$ are periodic functions of y_1 and y_2, where the period is 2π with respect to y_1 and 4π with respect to y_2. Furthermore, after a detailed analysis he showed that it was also possible to ensure that the functions remained finite and so could be expanded as sines and cosines of multiples of y_1 and $\dfrac{y_2}{2}$, where if p is even they will contain only the even multiples of $\dfrac{y_2}{2}$ and if p is odd they will contain only the odd multiples of $\dfrac{y_2}{2}$.

He was then able to write the approximate equations for the asymptotic surfaces as the asymptotic series

$$x_1 = \sum_{p=0}^{p=n} \mu^{\frac{p}{2}} \frac{\partial S_p}{\partial y_1}, \qquad x_2 = \sum_{p=0}^{p=n} \mu^{\frac{p}{2}} \frac{\partial S_p}{\partial y_2}.$$

These series, as he had previously proved, are divergent; but since they are asymptotic, if they are stopped at the nth term then the error is very small, providing, of course, μ is very small.

This was a crucially different conclusion from the mistaken one Poincaré had reached in *Note F*, where he had been misled by his false belief in the convergence of the series. There he had said that the equations for the asymptotic surfaces could be represented by the convergent infinite series

$$x_1 = \sum_{p=0}^{p=\infty} \mu^{\frac{p}{2}} \frac{\partial S_p}{\partial y_1}, \qquad x_2 = \sum_{p=0}^{p=\infty} \mu^{\frac{p}{2}} \frac{\partial S_p}{\partial y_2}.$$

5.8.3. Third approximation. In the final refinement Poincaré constructed exactly the asymptotic surfaces or, strictly speaking, their intersection with the transverse section $y_1 = 0$. Here the differences between [P1] and [P2] are quite dramatic.

In [P1] Poincaré's objective was to determine the coefficients of the series defining the asymptotic surfaces. He began by quickly disposing of the two cases where the series were clearly divergent. In the first case, when $C_1 > -\phi_4$, he likened the series to those derived by Lindstedt: divergent but nonetheless useful since the divergence derives from large multipliers rather than small divisors and so is relatively slow. In the second, when $C_1 < -\phi_4$, he gave an analysis which became the introduction to *Periodic solutions of the second class* in [P2] and is discussed later.

He moved on to the case where $C_1 = -\phi_4$, which he then believed gave rise to convergent series defining the asymptotic surfaces. He therefore set about determining the coefficients of the series given the properties he thought he had previously established, namely, that they were periodic with respect to y_1, that they were real and finite, and that they were convergent for sufficiently small values of μ. This

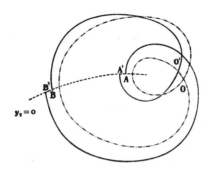

FIGURE 5.8.iii

involved showing it was possible to choose the series of constants derived in the section on the *Equation of Asymptotic Surfaces* so that the series were convergent. That being done, he returned to the geometry in order to give an actual description of the asymptotic surfaces.

To clarify the description, he used *Figure 5.8.iii*.[189] The plain lines which identify the two curves $AO'B'$ and $A'O'B$ represent the two asymptotic surfaces which cut the surface $y_1 = 0$, and the dashed line represents the curve $y_1 = y_2 = 0$. The dotted and dashed line, which is a closed curve with a double point at O, represents the curves with equation

$$y_1 = 0, \qquad \frac{x_2}{x_1} = \frac{x_2^0}{x_1^0} + \frac{\sqrt{\mu}}{x_1^0}\sqrt{\frac{2}{N}([F_1] - \phi_4)}$$

that arise when the surfaces which differ very little from the asymptotic surfaces cut the surface $y_1 = 0$. The generating (unstable) periodic solution is represented by a closed trajectory cutting the surface $y_1 = 0$ at the point O', and the distance OO' is of order μ.

Poincaré used his results from the end of the first approximation to infer, first, that the curve $BO'B'$ is quasi-closed, the distance between the points of closure being infinitely small of order μ, and, second, that the distance of the point B to its iterate is of the order of $\sqrt{\mu}$. Appealing to the (invalid) Corollary to Theorem III he concluded (erroneously) that the curve $BO'B'$ was rigorously closed, i.e., that the points B and B' were coincident, and consequently that the asymptotic surfaces were closed. Furthermore, inherent in this conclusion was the implication of stability.

He ended by noting that a similar argument could be used to establish that the asymptotic surfaces corresponding to the unstable periodic solution $C_1 = -\phi_2$ were also closed.

Thus Poincaré believed he had proved, for sufficiently small values of μ, that relative to a given unstable periodic solution there was a set of asymptotic solutions that could be considered stable in the sense that they remained confined to a given region of space. Moreover, he thought that this set of solutions was well behaved and that their behaviour could be completely understood. His analysis in [P2] led to an entirely different conclusion.

---

[189]Poincaré [P2, *438*].

In [P2] Poincaré used the same diagram as in [P1] (*Figure 5.8.iii*), with the same labelling, except that in [P2] the dotted and dashed line represents the curves with equation

$$y_1 = 0, \qquad x_1 = s_1^p(0, y_2), \qquad x_2 = s_2^p(0, y_2),$$

where s_i^p are the sums of the first p terms of the series $s_i(y_1, y_2)$ and are periodic functions of period 2π with respect to y_1 and 4π with respect to y_2.

The first question Poincaré asked was whether the curves $AO'B'$ and $A'O'B$ were closed. He could see that they would be if the series s_i were convergent. For in this case the plain curves would differ as little as required from the dotted and dashed curves, since the distance from a point on the former to a point on the latter would tend to zero as p increased indefinitely. But he had already proved that the series were divergent. Nevertheless, the question still remained. Was it possible for the curves $AO'B'$ and $A'O'B$ to be closed even though the series were divergent?

Poincaré tackled the question by looking at the specific example of a simple pendulum weakly coupled to a linear oscillator. In this case the Hamiltonian is given by

$$-F = p + q^2 - 2\mu \sin^2 \frac{y}{2} - \mu\epsilon \cos x \phi(y),$$

where μ and ϵ are two very small parameters, and $\phi(y)$ is a periodic function of y of period 2π. The Hamiltonian equations are

$$\frac{dp}{dt} = \frac{\partial F}{\partial x} = -\epsilon \sin x \phi(y), \qquad \frac{dq}{dt} = \frac{\partial F}{\partial y} = -\mu \sin y + \mu\epsilon \cos x \phi'(y),$$

$$\frac{dx}{dt} = -\frac{\partial F}{\partial p} = 1, \qquad \frac{dy}{dt} = -\frac{\partial F}{\partial q} = 2q,$$

where the variables p, q, x and y correspond to the variables x_1, x_2, y_1 and y_2 respectively in equations (27).

If $\epsilon = 0$, then the equations have a periodic solution

$$x = t, \qquad p = 0, \qquad q = 0, \qquad y = 0,$$

the two nonzero characteristic exponents are equal to $\pm \sqrt{2\mu}$, and the equations of the two asymptotic surfaces are

$$p = \frac{\partial S_0}{\partial x}, \qquad q = \frac{\partial S_0}{\partial y}, \qquad S_0 = \pm 2\sqrt{2\mu} \cos \frac{y}{2};$$

from which

$$p = 0, \qquad q = \pm\sqrt{2\mu} \sin \frac{y}{2},$$

and hence the surfaces enclose a region which has a width of the order of $\sqrt{\mu}$.

Since there are nonzero characteristic exponents for $\epsilon = 0$, there are also periodic solutions for small values of ϵ. The equations of the corresponding asymptotic surfaces are given by

$$p = \frac{\partial S}{\partial x}, \qquad q = \frac{\partial S}{\partial y},$$

where S is a function of x and y, satisfying the equation

$$\frac{\partial S}{\partial x} + \left(\frac{\partial S}{\partial y}\right)^2 = 2\mu \sin^2 \frac{y}{2} + \mu\epsilon \cos x \phi(y).$$

Moreover, the existence of nonzero characteristic exponents for $\epsilon = 0$ implies that p and q and therefore S can be expanded as series in ϵ. So S can be put in the form $S = S_0 + \epsilon S_1 + \epsilon^2 S_2 + \ldots$. S_0 has already been found, and equating powers of ϵ shows that S_1 must satisfy the equation[190]

$$\frac{\partial S_1}{\partial x} + 2\sqrt{2\mu} \sin\frac{y}{2}\frac{\partial S_1}{\partial y} = \mu \cos x \phi(y).$$

To determine S_1, Poincaré defined a new function $\sum$ to be the function which satisfies

$$\frac{\partial \sum}{\partial x} + 2\sqrt{2\mu} \sin\frac{y}{2}\frac{\partial \sum}{\partial y} = \mu e^{ix}\phi(y), \qquad i = \sqrt{-1},$$

so that S_1 is the real part of $\sum$. This equation can then be satisfied by $\sum = e^{ix}\Psi(y)$, which gives a linear equation in Ψ:

$$i\Psi + 2\sqrt{2\mu}\sin\frac{y}{2}\frac{d\Psi}{dy} = \mu\phi(y).$$

If $\phi = 0$, then[191]

$$\Psi = \left(\tan\frac{y}{4}\right)^\alpha, \qquad \alpha = -i\frac{1}{\sqrt{2\mu}},$$

and if $\phi(y)$ is arbitrary, the integral can be written

$$\Psi = \left(\tan\frac{y}{4}\right)^\alpha \int \sqrt{\frac{\mu}{8}}\phi(y)\left(\sin\frac{y}{2}\right)^{-1}\left(\tan\frac{y}{4}\right)^\alpha dy,$$

where Ψ can be expanded in integer powers of y for small values of y. If $\phi(0) = 0$, then the integral is also equal to zero and hence its limits are 0 and y.

If the curves $AO'B'$ and $A'O'B$ are closed,[192] then the function S and its derivatives will be finite for all values of y as well as being periodic of period 4π with respect to y (cf. the functions s_1^p and s_2^p). Since this must be true for any given value of ϵ, it must also be true for S_1, and hence for Ψ.

Thus, for values of y close to 2π, Ψ should be expansible in integer powers of $y - 2\pi$. But since $\left(\tan\frac{y}{4}\right)^\alpha$ cannot be expanded in this way, the condition can only hold if the integral

$$J = \int_0^{2\pi} \sqrt{\frac{\mu}{8}}\phi(y)\left(\sin\frac{y}{2}\right)^{-1}\left(\tan\frac{y}{4}\right)^{-\alpha} dy$$

is zero. However, evaluating J, using $\phi(y) = \sin y$, gives

$$J = -2\pi i \operatorname{sech}\left(\frac{\pi}{2\sqrt{2\mu}}\right),$$

which is clearly not equal to zero, and so the curves $AO'B'$ and $A'O'B$ cannot be closed.

However, the lack of closure still left open the possibility that the extended curves $O'B$ and $O'B'$ could intersect. For if this should occur, any trajectory which passes through the point of intersection would simultaneously belong to both sides

---

[190]Poincaré omitted the factor 2 in the second term of this equation, although he included it in [P1a] and then crossed it out.

[191]Poincaré put $\alpha = -i\sqrt{\frac{2}{\mu}}$.

[192]Poincaré wrote BO'B' and AO'A'.

of the asymptotic surface. To distinguish this type of trajectory, Poincaré called them *doubly asymptotic*. He was later to use the term *homoclinic*.[193]

In other words, if C is the closed trajectory which passes through the point O' and represents the periodic solution, then, if the trajectory is doubly asymptotic, it would begin by being very close to C when t is very large and negative. It would then asymptotically move away to deviate greatly from C, before asymptotically reapproaching C when t is very large and positive.

To prove the existence of doubly asymptotic trajectories Poincaré needed to show that the system fulfilled the conditions of Theorem III in Chapter III of Part 1.

To do this he established that none of the curves $O'B$, $O'B'$, $O'A$ and $O'A'$ were self-intersecting, i.e., that none of them have a double point; that the curvature of the curves $O'B$ and $O'B'$ was finite, i.e., that it does not increase indefinitely as μ tends to zero; and that the distances BB', B_1B_1', together with the ratios $\dfrac{BB'}{BB_1}$ and $\dfrac{BB'}{B'B_1'}$, tend to zero as μ tends to zero, where B_1 and B_1' are the iterates of B and B' respectively. Furthermore, since the system is in Hamiltonian form, it also possesses a positive invariant integral, and hence all the conditions of Theorem III are satisfied.

Therefore the arcs BB_1 and $B'B_1'$ intersect each other, i.e., the extended curve $O'B'$ intersects the extended curve $O'B$, and through the point of intersection (today called a *homoclinic* point) passes a doubly asymptotic trajectory.

Poincaré had constructed the figure so that the points B and B' lie on the curve
$$y_1 = y_2 = 0,$$
and since the origin of y_2 is arbitrary he supposed that at the intersection of the curves $O'B$ and $O'B'$, $y_2 = 0$. In this case the points B and B' are coincident and so are their iterates B_1 and B_1'. Thus the two arcs BB_1 and $B'B_1'$ have the same end points. But by Theorem III (in which the area limited by the two arcs is not convex), the two arcs must intersect again at a different point N. Thus there are at least two doubly asymptotic trajectories, one passing through the point B and one passing through the point N.

To show that there are in fact an infinite number of doubly asymptotic trajectories, he supposed that the points B and B' are always coincident, that BMN is the part of the curve $O'B$ between B and N; and BPN is the part of the curve $O'B'$ between the point $B = B'$ and the point N, so that these two arcs limit a certain area α. If the system is one in which the conditions of the recurrence theorem are satisfied, such as the restricted three body problem, then there will be trajectories which cross this area α infinitely often. Hence among the iterates of the area α there will be an infinite number which have a part in common with α.

The closed curve $BMNPB$ which limits the area α has an infinite number of iterates. The arc BMN cannot intersect any of its own iterates; for the arc BMN and its iterates belong to the curve $O'B$ and the curve $O'B$ is not self-intersecting.

---

[193]Poincaré first used the word 'homocline' in [MN III, *384*].

Similarly, the arc BPN does not intersect any of its own iterates. Therefore either the arc BMN intersects with one of the iterates of BPN, or the arc BPN intersects with one of the iterates of BMN (as in the case under consideration). In either case the curve $O'B$ or its extension will intersect the curve $O'B'$ or its extension.

Thus these two curves intersect each other at an infinite number of points, and an infinite number of these points of intersection will be found either on the arc BMN or on the arc BPN. These points of intersection are all points of intersection of the curve $O'B'$ or its extension with the curve $O'B$ or its extension, and, since through each of these points of intersection passes a doubly asymptotic trajectory, there are an infinite number of doubly asymptotic trajectories. Similarly, the asymptotic surface which cuts the surface $y_1 = 0$ along the curve $O'A$ also contains an infinite number of doubly asymptotic trajectories.

This is arguably the first mathematical description of chaotic motion within a dynamical system. Although Poincaré drew little attention to the complexity of the behaviour he had discovered and made no attempt to draw a diagram, he was profoundly disturbed by his discovery, as he revealed in a letter to Mittag-Leffler:

> I have written this morning to M. Phragmén to tell him of an error I have made and doubtless he has shown you my letter. But the consequences of this error are more serious than I first thought. It is not true that the asymptotic surfaces are closed, at least in the sense which I originally intended. What is true is that if both sides of this surface are considered (which I still believe are <u>connected</u> to each other) they intersect along an infinite number of asymptotic trajectories (and moreover that their distance becomes infinitely small of order higher than μ^p however great the order of p).
>
> I had thought that <u>all</u> these asymptotic curves having moved away from a closed curve representing a periodic solution, would then asymptotically approach the <u>same</u> closed curve. What is true, is that there are an infinity which enjoy this property.
>
> I will not conceal from you the distress this discovery has caused me. In the first place, I do not know if you will still think that the results which remain, namely the existence of periodic solutions, the asymptotic solutions, the theory of characteristic exponents, the nonexistence of single-valued integrals, and the divergence of Lindstedt's series, deserve the great reward you have given them.
>
> On the other hand, many changes have become necessary and I do not know if you can begin to print the memoir; I have telegraphed Phragmén.
>
> In any case, I can do no more than to confess my confusion to a friend as loyal as you. I will write to you at length when I can see things more clearly.[194]

---

[194]Poincaré to Mittag-Leffler, postmarked 1.12.1889, No. 54a, I M-L. (The original letter is reproduced on the following pages.)

††. et de plus que leur distance est un infiniment petit d'ordre plus élevé que p^i quelque grand que soit p.

[54 a]

Mon cher ami,

J'ai écrit ce matin à M. Phragmén pour lui parler d'une erreur que j'avais commise et il vous a sans doute communiqué ma lettre. Mais les conséquences de cette erreur sont plus graves que je ne l'avais cru d'abord. Il n'est pas vrai que les surfaces asymptotiques soient fermées, au moins dans le sens où je l'entendais d'abord. Ce qui est vrai, c'est que si l'on considère les deux parties de cette surface (que je croyais liées encore raccordées l'une à l'autre) se coupent suivant une infinité de courbes trajectoires asymptotiques. ††.

J'avais cru que toutes ces courbes asymptotiques après s'être éloignées d'une courbe fermée représentant une solution périodique, se rapprochaient ensuite asymptotiquement de la même courbe fermée. Ce qui est vrai, c'est qu'il y en a une infinité qui jouissent de cette propriété.

Je ne vous dissimulerai pas le chagrin que me cause cette découverte. Je ne sais d'abord si vous jugerez encore que les résultats qui subsistent, à savoir l'existence des solutions périodiques, celle des solutions asymptotiques; la théorie des exposants caractéristiques, la non-existence des intégrales uniformes, et la divergence des séries de M Lindstedt méritent encore la haute récompense que vous avez bien voulu leur accorder.

D'autre part, de grands remaniements vont devenir nécessaires et je ne sais si on n'a pas commencé à tirer le mémoire qu'si tu télégraphie à M. Phragmén,

En tout cas je ne puis mieux faire que de confier mes perplexités à votre ami aussi dévoué que vous l'avez toujours été.

Je vous en écrirai plus long quand j'aurai vu un peu plus clair dans mes affaires.

Veuillez agréer, mon cher ami, avec mes bien sincères excuses, l'assurance de mon entier dévouement,

Poincaré

Perhaps a further indication of Poincaré's concern and confusion at his discovery of the strange behaviour of these solutions can be detected in the introduction to [P2]. Of all the results in the memoir this was clearly the most extraordinary, and yet it is not amongst those he singled out in his introduction. Possibly this was because he felt unable to do so without ignoring Mittag-Leffler's request not to give details of the error. It would after all have been very difficult to draw attention to the complexity of the doubly asymptotic solutions without mentioning the error. On the other hand perhaps it was simply because he had had so little time in which to assess the implications of his discovery that he felt it wiser not to emphasise it.

5.9. Further results

The penultimate chapter of [P2] is devoted to three separate topics: periodic solutions of the second class, the divergence of Lindstedt's series and the nonexistence of any new integrals for the restricted three body problem.

Most of the chapter is derived from [P1] enhanced by additions. The first section on periodic solutions of the second class contains most of the section with the same name in [P1], although it opens with part of the section entitled *The Exact Construction of Asymptotic Surfaces* and concludes with some new material. The section on the divergence of Lindstedt's series is essentially *Note A*, while the final section on the nonexistence of single-valued integrals is derived from *Note G*.

The last two sections contain the results which Poincaré had emphasised in the introduction to [P2] and which quickly came to be amongst the best known in the memoir. Given the relative importance he now attached to them, it is perhaps surprising that he did not include their complete proofs in [P1]. It may well have been, of course, that they were not in a sufficiently finished form by the closing date of the competition. On the other hand, his emphasis on them in [P2] could possibly represent Weierstrass's opinion, which had been conveyed to him by Mittag-Leffler (see Chapter 6).

5.9.1. Periodic solutions of the second class.
In Poincaré's investigation of asymptotic solutions he had begun by showing how the periodic solutions could be represented by curves on the transverse section defined by the surface $y_1 = 0$, the nature of the curves depending on the value of a particular constant C_1, and his discussion had centred on the unstable periodic solution corresponding to $C_1 = \phi_4$. In this section he considers the situation when the value of this constant is less than $-\phi_4$, i.e., when the coefficient $x_2^1 = \sqrt{\frac{2}{N}([F_1] + C_1)}$ in the series for x_2 is not always real.

He found that in this case there were regions of motion but that these regions contained periodic solutions which made more than one revolution around the origin before closing up. These solutions were therefore of a different type from those that he had previously found, and he labelled them periodic solutions of the second class. More formally they can be described by saying that if a system has, for small values of μ, a periodic solution of period T, then they are those periodic solutions which are close to the original periodic solution but whose periods are integral multiples of T.

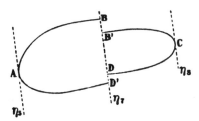

FIGURE 5.9.i

Since $[F_1]$ is a function of y_2, the behaviour of the system for different values of the constant C_1 will depend on y_2. If for a chosen value of C_1, x_2^1 is real as the value of y_2 varies between, say, η_5 and η_6, then since x_i^k are determined by a recurrence relation that is dependent on x_2^1, x_i^k can be determined for all values η_7 of y_2 in this range. The existence of the square root means that x_i^k has two sets of values equal in magnitude but opposite in sign. If $x_{0.i}^k$ are the functions of y_2 when the square root is positive and $x_{1.i}^k$ are the functions of y_2 when it is negative, then the latter will be the analytic continuation of the former.

What Poincaré now wanted to establish was how the behaviour of the system was affected by a change in the constant C_1. Were there regions in which the value of y_2 would remain finite? Since his method of using a transverse section to understand the evolution of the system reduced the dimension of the system by one, it was possible that a small change in the constant could induce some strange behaviour in y_2 which would only manifest itself in this other (unseen) dimension. It was clear that any behaviour of this type would not be captured by a study of the transverse section for individual choices of the constant.

Poincaré therefore looked at the change in values of y_2 for a very small change in the constant C_1. He replaced C_1 by a new constant C_1' very close to C_1, so that $\sqrt{\frac{2}{N}([F_1] + C_1')}$ was real whenever y_2 was between η_7 and a certain value η_8 very close to η_6. Again he could determine the functions x_i^k for all values of y_2 in this range, where $x_{2.i}^k$ are the functions of y_2 when the square root is positive and $x_{3.i}^k$ the functions of y_2 when it is negative.

He then constructed the four branches of the curve:

$$y_1 = 0, \quad x_1 = \phi_{k.1}(y_2), \quad x_2 = \phi_{k.2}(y_2),$$
$$(k = 0, 1 : \eta_5 \leq y_2 \leq \eta_7; \quad k = 2, 3 : \eta_7 \leq y_2 \leq \eta_8)$$

where

$$\phi_{p.q}(y_2) = x_{p.q}^0 + x_{p.q}^1 \sqrt{\mu} + \ldots + x_{p.q}^k \mu^{k/2},$$

the first and second branches of the curve corresponding to the constant C_1, and meeting at a tangent to the curve $y_2 = \eta_7$, and the third and fourth branches of the curve corresponding to the constant C_1' and meeting at a tangent to the curve $y_2 = \eta_8$ (*Figure 5.9.i*).[195]

Poincaré began by regarding C_1 as given and C_1' as close in value to C_1 but nevertheless arbitrary. He now determined C_1' by imposing the condition that the

---

[195]Poincaré [P2, *447*].

first and third curves meet, i.e., the points B and B' are coincident. Then appealing to an earlier theorem (invariant integrals, Theorem III), he derived the result that the distance DD' between the second and the fourth curves was infinitely small of the order $\mu^{(k+1)/2}$. Hence he could conclude that for a limited period of time there do exist regions in which the values of y_2 remain finite, and these are those known as the regions of libration. The time constraint was a consequence of the fact that the series involved were asymptotic rather than convergent.

Poincaré then considered the regions of libration in order to ascertain whether they contained periodic solutions.

The equations

$$(34) \qquad x_1 = x_1^0 + \mu x_1^2, \quad x_2 = x_2^0 + \sqrt{\mu}\sqrt{\frac{2}{N}([F_1] + C_1)} + \mu u_2^2$$

define, up to the order of μ, the surfaces just constructed (see *Figure 5.9.i*) and so approximately satisfy equations (30), where u_2^2 is a finite function of y_1 and y_2 which only differs from x_2^2 by a function of y_2 so that

$$\frac{\partial(u_2^2)}{\partial y_1} = \frac{\partial(x_2^2)}{\partial y_1}.$$

Poincaré then modified the form of F in the Hamiltonian equations so that the equations (34) *exactly* satisfied equations (30). The new form of the equations can then be integrated exactly, and, following the same argument as given previously, shows that there exist an infinite number of closed surface trajectories defined by the equation

$$(35) \qquad \frac{x_2}{x_1} = \frac{x_2^0 + \sqrt{\mu}\sqrt{\frac{2}{N}([F_1] + C_1)} + \mu u_2^2}{x_1^0 + \mu x_1^2},$$

which is of the same form as equation (33).

Thus the same hypotheses can be made about the function $[F_1]$ in equation (35) as about $[F_1]$ in equation (33). The two surfaces of (35) which correspond to the values $-\phi_2$ and $-\phi_4$ of C_1 are therefore closed with a double curve. The space can then be divided into three regions: interior, exterior, and between the two sheets of the surface, the last being the region of libration.

Since the closed surface corresponding to a given value of C_1 (which must be $< -\phi_4$) has the same connectivity as a torus, the existence of periodic solutions depends on the behaviour of the two angular variables defining the surface. By investigating the behaviour of these angular variables Poincaré showed that there were an infinite number of values for C_1 for which periodic solutions exist.

Then, by deriving an equation in which C_1 was defined as a continuous function of μ, he showed that it was possible to make μ so small that the equation no longer had a root. This means that while there exist an infinite number of closed trajectories which represent periodic solutions, these solutions will disappear one after the other as μ decreases. In other words, if along a closed trajectory, μ decreases continuously, then the trajectory deforms continuously and at a certain value of μ will eventually disappear. As a result, when $\mu = 0$ all the periodic solutions in the region of libration will have disappeared. This is in contrast to the behaviour of the periodic solutions of the first class (those which only have one revolution around the primary) which continue to exist for $\mu = 0$.

Since he had proved that in the neighbourhood of a periodic solution (stable or unstable) there exist an infinite number of other periodic solutions, Poincaré considered the possibility that every region of the space, however small, was crossed by an infinite number of periodic solutions. In other words he conjectured that the periodic solutions were everywhere dense. Although he was unable to prove it, Cantor's recent work on set theory, which had shown that it was possible for a set to be perfect without being connected, led him to believe that it was extremely likely.[196]

All the above is contained in both [P1] and [P2]. There is, however, an important addition to the section which only appears in [P2]. In both versions of his proof of the existence of periodic solutions of the second class, Poincaré had shown that periodic solutions exist for small values of a certain parameter ϵ. But in [P1] he had thought that if $\epsilon = \mu\sqrt{\mu}$ it automatically *followed* that periodic solutions also exist for small values of the mass parameter μ. In the revision he realised that this result needed to be established rigorously.

Therefore, in [P2] he returned to the Hamiltonian equations (27) and considered a *stable* periodic solution

$$x_1 = \phi_1(y_1), \qquad y_2 = \phi_2(y_1),$$

of period 2π. In *Figure 5.8.ii* this periodic solution is approximately represented by the isolated closed curve of the surface $C_1 = -\phi_3$. It has two characteristic exponents $\pm\alpha$, the squares of which are real and negative. If

$$x_1 = \phi_1(y_1) + \xi_1, \qquad y_2 = \phi_2(y_1) + \xi_2$$

is a nearby solution and β_1 and β_2 are the initial values of ξ_1 and ξ_2 for $y_1 = 0$, and $\beta_1 + \Psi_1$ and $\beta_2 + \Psi_2$ are the values of ξ_1 and ξ_2 for $y_1 = 2k\pi$ (k an integer), then the solution will be periodic of period 2π if

(36) $$\Psi_1 = \Psi_2 = 0,$$

where Ψ_1 and Ψ_2 can be expanded in powers of β_1 and β_2 which depend on μ.

If β_1, β_2 and μ are regarded as the coordinates of a point in space, then the equations (36) represent a curve, each point of which corresponds to a periodic solution. If $\xi_1 = \xi_2 = 0$, then $\beta_1 = \beta_2 = 0$, which implies $\Psi_1 = \Psi_2 = 0$, and a periodic solution of period 2π is obtained which can also be regarded as being periodic with period $2k\pi$.

Thus the curve (36) consists of the entire μ axis. Poincaré proved that if $k\alpha$ is a multiple of $2i\pi$ when $\mu = \mu_0$, then there exists another branch of the curve (36) which passes through the point

$$\mu = \mu_0, \qquad \beta = 0, \qquad \beta_2 = 0,$$

and so, from the previous theory, for values of μ close to μ_0, there exist periodic solutions other than $\xi_1 = \xi_2 = 0$.

---

[196]This point is taken up again in [MN I, *82*]. See Chapter 7.

The proof, which depended on the theory of invariant integrals, involved expanding Ψ_1 and Ψ_2 in terms of β_1, β_2 and $(\mu - \mu_0)$, and then showing that β_1 and β_2 could themselves be expanded in positive fractional powers of $(\mu - \mu_0)$. This implies that there exists a series in $(\mu - \mu_0)$ which is not identically zero and which satisfies equations (36). This in turn implies that there exists a system of periodic solutions in which the expressions for the coordinates can be expanded in positive fractional powers of $(\mu - \mu_0)$, the period of which is a multiple of the generating periodic solution, and when $\mu = \mu_0$ the solution is simply the original periodic solution.

5.9.2. Divergence of Lindstedt's series. In [P1] Poincaré had included a section entitled *Negative Results*, in which he had incorporated the result that no analytic single-valued integral of the restricted problem exists apart from the Jacobian integral. He claimed that a consequence of this result was that the series generally used in celestial mechanics, and in particular the series derived by Lindstedt (see Chapter 2), were, contrary to what had been previously thought, divergent. But he gave no evidence of why this assertion should be true. That it was not immediately obvious is plainly expressed by Mittag-Leffler, who told Poincaré that he had spent a month with Weierstrass trying to work it out but without success.[197] Poincaré responded with *Note A*, in which he gave two forms of a proof, but, according to Mittag-Leffler, Weierstrass was still not satisfied. Poincaré had proved that there were circumstances under which Lindstedt's series were not convergent, but he had ignored the wider question of whether convergent trigonometric series solutions could ever be found. Furthermore, since Dirichlet's original remarks had led Weierstrass to believe that such solutions did exist, he was particularly anxious to have this important point clarified.[198] Mittag-Leffler again asked Poincaré for a proof.[199] This time Poincaré replied that he thought that he had covered the point in *Note A*, although he admitted that he could not be sure as he had mislaid his own copy of it.[200] Mittag-Leffler did not pursue the issue, in spite of further requests from Weierstrass to do so, and, significantly, Poincaré left Weierstrass's question unresolved. In [P2] Poincaré gave only the first form of the proof, since the second depended on the proof of the nonexistence of any new integrals which he had placed in the following section.

Poincaré first showed how Lindstedt's method for approximately integrating differential equations of the form

$$\frac{d^2x}{dt^2} + n^2x = \alpha\Phi(x,t)$$

by deriving a formal trigonometric series solution without any secular terms could be adapted to accommodate the system of Hamiltonian equations (27). He considered the system with Hamiltonian

$$F = F_0 + \mu F_1,$$

where F_0 is independent of y_1 and y_2, and F_1 is a trigonometric series of sines and cosines of multiples of y_1 and y_2 with coefficients which are analytic functions of x_1

---

[197]Mittag-Leffler to Poincaré, 15.11.1888, I M-L.
[198]See Appendix 2, Question 1.
[199]Mittag-Leffler to Poincaré, 23.2.1889, I M-L.
[200]Poincaré to Mittag-Leffler, 1.3.1889, No. 49, I M-L.

and x_2. The x_i and y_i are then regarded as functions of two variables $w_i = \lambda_i t + \varpi_i$ (as opposed to simply functions of the time), where the frequencies λ_i are to be determined, ϖ_i are constants of integration, and

$$(37) \qquad x_i = x_i^0 + \mu x_i^1 + \mu^2 x_i^2 + \ldots \qquad (i = 1, 2)$$
$$y_i = w_i + \mu y_i^1 + \mu^2 y_i^2 + \ldots$$
$$\lambda_i = \lambda_i^0 + \mu \lambda_i^1 + \mu^2 \lambda_i^2 + \ldots .$$

The coefficients λ_i^k are constants, and the coefficients x_i^k and y_i^k are trigonometric series in sines and cosines of multiples of w_1 and w_2. Poincaré then sketched a method which, in line with Lindstedt's result, demonstrated that it was possible to determine the $2q + 2$ constants $\lambda_i^0, \ldots, \lambda_i^q$ so that the $4q$ trigonometric series x_i^0, $\ldots, x_i^q, y_i^0, \ldots, y_i^q$, for arbitrarily large q, satisfy the Hamiltonian equations up to the order of μ^{q+1}.

The frequencies λ_1 and λ_2 can be expanded in powers of μ, ω_1 and ω_2, and the solutions corresponding to the values of ω_1 and ω_2 for which the frequency ratio is commensurable are therefore periodic. Corresponding to each of these periodic solutions are characteristic exponents each of which can be calculated if the general solution of the equations is known. Thus if the series are uniformly convergent and consequently give the general solution of the equations, then the characteristic exponents of the periodic solutions can be calculated.

When Poincaré calculated the characteristic exponents under the assumption that analytic solutions do exist he found that they were all zero. But when he put this result into the eigenvalue equation that determines the characteristic exponents in the restricted problem, he arrived at a contradiction. He therefore concluded that his original assumption—that there is a general solution given by uniformly convergent series—must be false, and hence:

> ... in the restricted three body problem and consequently in the general
> case, Lindstedt's series are not uniformly convergent for all the values
> of the arbitrary constants of integration which they contain.[201]

But, as Weierstrass had observed, Poincaré's discussion was incomplete. He gave no consideration to the circumstances under which convergence could occur, with the result that he gave no indication of what proportion of the series were divergent. However, Poincaré did not abandon the question. In the second volume of the *Méthodes Nouvelles* he reworked it in greater depth and generality, and his conclusions are described in Chapter 7 below.

5.9.3. Nonexistence of single-valued integrals.

The final section in the chapter contained what has become one of the best-known results in the memoir: the proof of the nonexistence of any new transcendental integral for the restricted three body problem. Only two years earlier Heinrich Bruns, a former student of Weierstrass, had proved the nonexistence of any new algebraic integral for the general three body problem [1887],[202] and Poincaré's result was therefore an important complement to that of Bruns'.

---

[201] [P2, *470*].

[202] For a clear exposition of Bruns' result see Whittaker [1937, *358*].

Poincaré had given an outline of a proof of this result in the *Negative Results* in [P1], and in response to yet another of Mittag-Leffler's requests for details, he had also produced an extension to the proof which he had intended to appear as *Note G.*[203] In [P2], having reshaped his original thoughts, he completely rewrote the proof.

More specifically, Poincaré proved that if the differential equations (27) possess the solution $F = constant$, where F is a single-valued analytic function of x_i, y_i and μ, which can be expanded in powers of μ and is periodic of period 2π with respect to y_i, then the equations possess no other solution of the same form.

Suppose $\Phi = constant$ is another such solution. If $x_1 = \phi_1$, $x_2 = \phi_2$, $y_1 = \phi_3$, $y_2 = \phi_4$ is a periodic solution of the differential equations such that $x_1 = \phi_1 + \beta_1$, ..., $y_2 = \phi_4 + \beta_4$ when $t = 0$ and $x_1 = \phi_1 + \beta_1 + \Psi_1$, ..., $y_2 = \phi_4 + \beta_4 + \Psi_4$, when $t = T$, then Ψ can be expanded as a power series in β, and the eigenvalue equation in S

$$
\begin{vmatrix}
\dfrac{\partial \Psi_1}{\partial \beta_1} - S & \dfrac{\partial \Psi_1}{\partial \beta_2} & \dfrac{\partial \Psi_1}{\partial \beta_3} & \dfrac{\partial \Psi_1}{\partial \beta_4} \\[2ex]
\dfrac{\partial \Psi_2}{\partial \beta_1} & \dfrac{\partial \Psi_2}{\partial \beta_2} - S & \dfrac{\partial \Psi_2}{\partial \beta_3} & \dfrac{\partial \Psi_2}{\partial \beta_4} \\[2ex]
\dfrac{\partial \Psi_3}{\partial \beta_1} & \dfrac{\partial \Psi_3}{\partial \beta_2} & \dfrac{\partial \Psi_3}{\partial \beta_3} - S & \dfrac{\partial \Psi_3}{\partial \beta_4} \\[2ex]
\dfrac{\partial \Psi_4}{\partial \beta_1} & \dfrac{\partial \Psi_4}{\partial \beta_2} & \dfrac{\partial \Psi_4}{\partial \beta_3} & \dfrac{\partial \Psi_4}{\partial \beta_4} - S
\end{vmatrix} = 0
$$

can be formed.

The roots of this equation are $e^{\alpha T} - 1$, α being the characteristic exponents. Since it is the restricted three body problem that is being considered, two of the roots are zero and two are nonzero.

Furthermore,

$$
\frac{\partial F}{\partial x_1}\frac{\partial \Psi_1}{\partial \beta_i} + \frac{\partial F}{\partial x_2}\frac{\partial \Psi_2}{\partial \beta_i} + \frac{\partial F}{\partial y_1}\frac{\partial \Psi_3}{\partial \beta_i} + \frac{\partial F}{\partial y_2}\frac{\partial \Psi_4}{\partial \beta_i} = 0 \qquad (i = 1, \dots, 4)
$$

$$
\frac{\partial \Phi}{\partial x_1}\frac{\partial \Psi_1}{\partial \beta_i} + \frac{\partial \Phi}{\partial x_2}\frac{\partial \Psi_2}{\partial \beta_i} + \frac{\partial \Phi}{\partial y_1}\frac{\partial \Psi_3}{\partial \beta_i} + \frac{\partial \Phi}{\partial y_2}\frac{\partial \Psi_4}{\partial \beta_i} = 0,
$$

where, in the derivatives of F and Φ, x_i and y_i are replaced by $\phi_i(T)(i = 1, \dots, 4)$. Hence either

$$
(38) \qquad \frac{\partial F}{\partial x_1} \bigg/ \frac{\partial \Phi}{\partial x_1} = \frac{\partial F}{\partial x_2} \bigg/ \frac{\partial \Phi}{\partial x_2} = \frac{\partial f}{\partial y_1} \bigg/ \frac{\partial \Phi}{\partial y_1} = \frac{\partial F}{\partial y_2} \bigg/ \frac{\partial \Phi}{\partial y_2}
$$

or the Jacobian of Ψ with respect to β is zero, together with all the minors of first order.

---

[203]Mittag-Leffler to Poincaré 21.12.1888, I M-L.

On the other hand, if $\phi'(t)$ is the derivative of $\phi(t)$, then

$$\sum \frac{\partial \Psi_i}{\partial \beta_j} \phi'_j(0) = 0 \qquad (i = 1, \dots, 4)$$

$$\frac{\partial F}{\partial x_1} \phi'_1(0) + \frac{\partial F}{\partial x_2} \phi'_2(0) + \frac{\partial F}{\partial y_1} \phi'_3(0) + \frac{\partial F}{\partial y_2} \phi'_4(0) = 0$$

$$\frac{\partial \Phi}{\partial x_1} \phi'_1(0) + \frac{\partial \Phi}{\partial x_2} \phi'_2(0) + \frac{\partial \Phi}{\partial y_1} \phi'_3(0) + \frac{\partial \Phi}{\partial y_2} \phi'_4(0) = 0,$$

and if it can be shown that if equations (38) are not satisfied, then either

$$(39) \qquad \phi'_1(0) = \phi'_2(0) = \phi'_3(0) = \phi'_4(0) = 0$$

or the equation in S has three zero roots. But since S only has two zero roots and equations (39) are only satisfied for certain particular periodic solutions where the planetoid has a circular orbit,[204] the equations (38) must be satisfied for $x_1 = \phi_1(T)$, .., $y_2 = \phi_4(T)$. Furthermore, since the origin of the time is arbitrary, they must also be satisfied for $x_1 = \phi_1(t)$, .., $y_2 = \phi_4(t)$, that is for all points of the periodic solutions.

For the final part of the proof Poincaré showed that equations (38) were in fact satisfied identically, which proves that Φ is a function of F. Hence the two solutions F and Φ cannot be distinct. Thus the equations (37) do not admit any new single-valued transcendental integral, providing the value of μ is sufficiently small.

5.9.4. Positive and negative results. In [P1] Poincaré did not have a chapter dealing specifically with any of the three topics described above, that is, periodic solutions of the second class, the divergence of Lindstedt's series, and the nonexistence of integrals; although, as has been described, much of the material did appear in [P1] in a less organised fashion. However, he did include a chapter in [P1] in which he gave a general resumé of his results. The chapter was brief, amounting to only five pages, and was divided into two parts, positive and negative results, of which very little was reproduced in [P2].

With regard to the positive results Poincaré mainly concluded that the trajectories in the restricted three body problem could be classified into three types: closed trajectories corresponding to periodic solutions; asymptotic trajectories; and the general trajectories that did not fit into either of the above categories. He commented that the difficult and unexpected result which (he believed) he had established was that the trajectories which asymptotically approach an unstable closed trajectory were the same trajectories as those which asymptotically move away from the same unstable closed trajectory, and that the set of these asymptotic trajectories formed a closed asymptotic surface. It was of course the latter part of this result which was erroneous and which caused him so much trouble.

Of the results he described as negative, although he mentioned the divergence of Lindstedt's series, the one he considered the most important was the one concerning the nonexistence of any new integrals. He gave only a brief outline of the proof of this result in which he related it to the restricted problem, although, as mentioned, he did add a more detailed proof in *Note G*.

---

[204]See Laplace [1789-1825, **X**, Chapter VI].

5.10. Attempts at generalisation

5.10.1. The n body problem. In the last chapter of the memoir Poincaré returned to the competition question: the n body problem. Initially he had hoped in some way to generalise his earlier results, even allowing for the fact that there would be complications arising from the increased number of variables and the consequent impossibility of creating a realisable geometric representation. Unfortunately, this did not turn out to be the case, although the difficulties only really became apparent when he attempted to generalise the second half of the memoir.

The first part of the memoir presented him with little problem, since it was only a question of extending the number of dimensions of the representation space. This is quite straightforward, since in a system with two degrees of freedom a state of the system is represented by the position of a point in a space of 3 dimensions, while in a system with p degrees of freedom a state is represented by the position of a point in a space of $2p-1$ dimensions. Thus many of the conclusions concerning periodic and asymptotic solutions of the restricted problem can be generalised with relative ease. For example, the theory extends in a natural way to show that in the n body problem there are an infinite number of periodic solutions, stable and unstable, as well as an infinite number of asymptotic solutions.

It was in his attempt to generalise the second part of the memoir that Poincaré found himself beset with difficulties. He cited, for example, the case where the autonomous Hamiltonian equations have three degrees of freedom, and the problem is then to find three functions, $x_i = \Phi_i(y_1, y_2, y_3)$, satisfying

$$\frac{\partial x_i}{\partial y_1}\frac{\partial F}{\partial x_1} + \frac{\partial x_i}{\partial y_2}\frac{\partial F}{\partial x_2} + \frac{\partial x_i}{\partial y_3}\frac{\partial F}{\partial x_3} + \frac{\partial F}{\partial y_i} = 0 \qquad (i=1,2,3).$$

He found that even this relatively simple case led to the consideration of three different situations, two of which led to the problem of small divisors, while the third led to inscrutable integrals.

A second difficulty he faced concerned the motion of the perihelions. In the unperturbed case when the system is in a state of Keplerian motion, since the Hessian of F_0 with respect to the linear variables x_i is zero, as well as the Hessian of any arbitrary function of F_0, the perihelions remain fixed. This difficulty does not arise in the restricted problem because it is not necessary to use the longitude of the perihelion, g, as a variable, since the variable $(g-t)$ can be chosen instead.

Poincaré also drew attention to the fact that he had not made a full investigation of the periodic solutions of the unperturbed motion in the three body problem, that is, when the orbits of the two smaller bodies or planets reduce to Keplerian ellipses. In his analysis he had only considered the obvious case of periodic solutions that arise when the two mean motions are commensurable, and he had not considered the possibility of any others. Thinking about this particular question led him to another idea: that of periodic motion resulting from two planets passing infinitely close to each other without actually colliding. It occurred to him that if the planets did move in such a way, then this would give rise to a change in their orbits which would give the appearance of a collision. He thought then that it might be possible to choose the initial conditions in such a way that these "collisions" occurred periodically. If this were the case, then discontinuous solutions would be obtained which would be proper periodic solutions of the Keplerian motion. He did not have time at this stage to pursue the idea further, although he did discuss it some ten years later [MN III, Chapter XXXI], when he called these solutions *periodic solutions of the second species.*[205]

Although it was evident that some of the difficulties Poincaré had encountered in trying to generalise his results would be overcome in the fulness of time, there were others that appeared to be beyond the scope of available techniques. In any event, Poincaré had made it clear that the n body problem was still far from being solved.

With characteristic modesty, Poincaré concluded by saying that he regarded his work as only a preliminary survey from which he hoped future progress would result.

---

[205]These solutions are discussed further in Chapter 7.

CHAPTER 6

Reception of Poincaré's Memoir

6.1. Introduction

The discovery of the error in Poincaré's original memoir and the accompanying delay in the publication meant that almost two and a half years had elapsed between the submission of the manuscript to the competition and the publication of the corrected memoir in *Acta*. As a result it is possible to distinguish between the responses to each version. The response to [P1] consisted of the opinions of the prize commission, together with those whose knowledge of it was derived from Mittag-Leffler's report. [P2] had a rather wider audience.

At the time the prize was awarded the only publicly available information about the mathematical content of the prize-winning memoir was Mittag-Leffler's brief report. The memoir had not been made publicly available, and so the rest of the press reports covering Poincaré's triumph contained no mathematical commentaries. Furthermore, the prepublication copies of [P1] which made a brief appearance at the end of 1889 were in circulation for such a short period that none of the recipients would have had time to master the contents in order to make a fully informed judgement. Thus the only people who had had the opportunity to scrutinize [P1] were the members of the prize commission. Nevertheless, despite the paucity of information available in Mittag-Leffler's report, it was sufficient to provoke an adverse reaction from Hugo Gyldén.

When [P2] finally appeared, it drew widespread praise, although it is clear that certain aspects of Poincaré's mathematics were well beyond the grasp of his fellow mathematicians. With regard to the error, Mittag-Leffler's campaign of secrecy plus the delay caused by the backlog of publishing at *Acta* meant that when [P2] was published, those members of the mathematical community who had heard any rumours about the error had had plenty of time to forget the details. The result was that by the time [P2] finally appeared, any remaining concerns about the error, if indeed they existed, seemed to have vanished, despite Poincaré's brief allusion to it in the introduction.

6.2. The views of the prize commission

As described in Chapter 4 above, the three members of the prize commission were quick to come to a unanimous decision concerning the overall merit of Poincaré's entry. However, in their correspondence during the adjudication period, the only one who ventured anything more than a general opinion on the mathematics was Weierstrass. Mittag-Leffler, although he openly indicated to Poincaré the points in the memoir which he felt needed further elaboration, fought shy of discussing the relative merits of any of the results with either Poincaré or the members of the commission. His prime concern appears to have been in fulfilling his role as a mediator between Hermite and Weierstrass, communicating information from one to the other.

Of a more public nature was Mittag-Leffler's responsibility for the commission's general report. But since the report was intended to present the opinion of the whole commission on the results of the competition and was written with help from Hermite and Weierstrass, it gives no insight into Mittag-Leffler's personal views. In any case, since the task of providing a mathematical analysis of Poincaré's memoir had been left to Weierstrass, the report contained no details of the winning entry beyond giving a general indication of the nature of the results and emphasising the power of analytic methods in treating questions of celestial mechanics.[206]

In addition to writing the general report, Mittag-Leffler had been asked by the King to give a resumé of Poincaré's results at the February meeting of the Swedish Academy of Sciences. As it happened, Mittag-Leffler's talk was not given until March, and then in rather less than favourable circumstances, the background to which is described later in the chapter.

Hermite, by virtue of being in Paris, was in the unique position of being able to speak directly to Poincaré about the memoir. He was unequivocal in his opinion of it, but, as he told Mittag-Leffler, he too had sought help from Poincaré over the details:

> Poincaré's memoir is of such rare depth and power of invention, it will certainly open a new scientific era from the point of view of analysis and its consequences for astronomy. But greatly extended explanations will be necessary and at the moment I am asking the distinguished author to enlighten me on several important points.[207]

It is not clear exactly which parts of [P1] Hermite felt needed explaining, nor did he give any indication of the results he considered the most important. The implication that Hermite felt uneasy about his ability fully to comprehend the mathematics is confirmed by his reaction to the suggestion that he might have to write the official report on the memoir. Although it was more or less understood that Weierstrass as proposer of the question ought to be the author of the report, Mittag-Leffler, as a result of Weierstrass's declining health, had expressed concern to Hermite about Weierstrass's fitness for the undertaking.[208] Unfortunately, Hermite's response was not what Mittag-Leffler was looking for:

> The task of writing the report falls by right to Weierstrass who proposed the question, who can with authority express reservations which would put me into an indescribably difficult position should I have to make them. Indeed what would be my position vis-à-vis Poincaré to whom I would have to appeal for explanations in order to understand the most important points of the memoir; I would no longer be in the

---

[206]See Appendix 4.

[207]Hermite to Mittag-Leffler, 17.10.1888, *Cahiers* **6** (1985), *146*.

[208]Mittag-Leffler to Hermite, 17.10.1888, I M-L.

role of judge, and I must tell you in all frankness, if I have to make the report, it would be the echo of what I had heard from listening to the author, with the intention of justifying my admiration for his genius. ... Besides, to satisfy the demand of public opinion, and taking into account the importance and seriousness of the announcement of the prize, do you think that it is advisable that the prize awarded to a Frenchman should rest on the report of a Frenchman who is his colleague and friend?[209]

Mittag-Leffler needed no further convincing, and the responsibility for the report remained with Weierstrass, who had also displayed doubts about Hermite's ability to deal with the mathematics unaided.

At the 1889 public meeting of the Paris Academy of Sciences, Hermite, in his official capacity as Vice President, prompted by Mittag-Leffler's letter to the Secretary, used the occasion to comment on the results of the competition and commend the contents of Poincaré's memoir. In particular he drew attention to Poincaré's discovery of the asymptotic character of the series used in celestial mechanics and ironically chose to describe this result in terms of Poincaré having discovered an error: *"The error having been recognised, it opens a new avenue in the study of the three body problem, and this is where Poincaré's talent is displayed with brilliance."*[210] He could have had no idea how prescient those words would turn out to be.

Weierstrass's opinion of [P1] is mostly revealed in three letters to Mittag-Leffler, parts of which were published in *Acta* in Mittag-Leffler's [1912] biography of Weierstrass. The most significant of these is the first, in which he gave his judgement of the competition entries. Although he commented on five of them, he wrote more than four times more on Poincaré's work than on the other four put together.[211]

Weierstrass considered the most important results to be what Poincaré had described as *negative results*, that is, the divergence of Lindstedt's series and the theorem on the nonexistence of single-valued integrals. Although he thought that these showed that an entirely new approach would be needed for the problem to be solved, he was still convinced that a solution existed. With regard to the positive results, he singled out Poincaré's discoveries in stability, invariant integrals, periodic solutions and asymptotic solutions as being especially notable, and was very enthusiastic about the treatment of the analytic solutions of algebraic differential equations. But it was not unadulterated praise. Weierstrass had also found the memoir extremely difficult to read, and he was concerned about its general lack of rigour.

In the second letter written some seven weeks later, although he was even more enthusiastic about the memoir than before, Weierstrass confessed that despite working hard on it he still had not been able to master it completely.[212] He now believed that the results on periodic solutions and the discovery of asymptotic motion were achievements of the highest importance, describing them as no less

---

[209]Hermite to Mittag-Leffler, 22.10.1888, *Cahiers* **6** (1985), *147-148*.

[210]Hermite [1889].

[211]Weierstrass to Mittag-Leffler, 15.11.1888, I M-L. Mittag-Leffler [1912, *50-52*].

[212]Weierstrass to Mittag-Leffler, 8.1.1889, I M-L. Mittag-Leffler [1912, *53-55*].

than epoch-making. On a critical note he was concerned about Poincaré's treatment of the stability question in the restricted three body problem. He queried the physical validity of Poincaré's definition of stability, which appeared to put an upper bound on the distance between two points without considering what would happen if two points became infinitely close. As he pointed out, if this should occur, it would inevitably affect the form of motion, and thus, he argued, a distinction ought to be drawn that would take this into account.

The manuscript of this letter also reveals a careful piece of editing by Mittag-Leffler. Tactfully omitted from the published version is Weierstrass's remark in which he confided to Mittag-Leffler that he thought Hermite must have had somebody to explain the memoir to him.

In the third letter Weierstrass explained why, contrary to what he had said in his previous letter, he was now satisfied that Poincaré's analysis did ensure that the planetoid could not come infinitely close to the other two bodies.[213] He had simply overlooked Poincaré's incorporation of the condition that the value of the constant C in the equation

$$\frac{1}{2a} + G + \mu F_1 = C$$

had to be essentially greater than $\frac{3}{2}$. This was also the letter in which, as mentioned in the previous chapter, he questioned Poincaré's claim that the nonexistence of any new single-valued solution necessarily implied the nonexistence of convergent trigonometric solutions.

Finally, he told Mittag-Leffler that, although he would definitely have the report finished by the end of the week, he was having difficulty with its introduction. This was because he believed it should begin with a justification of the question in order to counter the adverse criticism which it had been lodged against it. The criticism was on two fronts: there were those who claimed that the question as it stood was completely insoluble, while others censured the limitation induced by the assumption that a collision between two points can never take place. Weierstrass indicated to Mittag-Leffler his intended response to these accusations, but his real concern was how to condense into a few lines something which he felt warranted a long discussion. Clearly much of the criticism was due to Kronecker [1888], and, given the brittle nature of their relationship, Weierstrass was keen for his defence to be carefully drafted.

Despite his intentions, Weierstrass never finished the report, although he did complete the introduction, sending a copy to Mittag-Leffler in March 1889. The introduction threw no light on his judgement of the memoir, but it did explain why he had chosen to formulate the n body problem in the way that he had, as well as giving the criteria he had used in judging the entries which had attempted to solve it.[214]

In the question Weierstrass had asked for an expansion of the coordinates as infinite series of known functions of time which were uniformly convergent for *unbegrenzter Dauer* (= unlimited time), the implication being that these series did actually exist, and it was the phrase *unbegrenzter Dauer* that was the main cause

---

[213]Weierstrass to Mittag-Leffler, 2.2.1889, I M-L. Mittag-Leffler [1912, *55-58*].
[214]Mittag-Leffler [1912, *63-65*].

of the misunderstanding. It had been interpreted as meaning that Weierstrass required series which were uniformly convergent for infinite time, i.e., for $0 \leq t \leq \infty$, rather than, as he had intended, series which were uniformly convergent for a fixed value of time, however large, i.e., for $0 \leq t \leq a$ $(a < \infty)$. The distinction was critical because Weierstrass believed he had proved that such series did exist in the latter case—the proof being dependent on its ability to show that the distance between any two points can never become either infinitely small or infinitely large as time approaches a finite limit—but he had no proof for the former.

Showing that the distance between two points can never become infinitely small amounts to dealing with the possibility of collisions, both binary and multiple. Weierstrass's difficulty was that he thought that when he had originally set the question he had constructed a proof that overcame the problem of collisions, but he could not remember it. In his letter to Mittag-Leffler he gave an outline of a proof that dealt with binary collisions and stated a conjecture about triple collisions, but gave no consideration to collisions of greater multiplicity.

To deal with a binary collision he assumed that the time t was sufficiently close to the moment of collision t_0 so that the coordinates of all the points could be expanded in positive powers of $(t_0 - t)^{\frac{1}{3}}$, and the expressions would contain not $6n$ but $6n - 2$ arbitrary constants.[215]

With regard to triple collisions, Weierstrass claimed that it was easy to show that all three points can only collide when the three constants of angular momentum are simultaneously zero. With regard to the three body problem this was clearly an important result, but unfortunately Weierstrass did not give a proof, and Mittag-Leffler did not press him for one. It was not until the beginning of the next century that Sundman, unaware of Weierstrass's conjecture, provided a proof of this result; this is discussed in Chapter 8.

Rather curiously Weierstrass does not appear to have considered the possibility of the mutual distances becoming infinitely large, and it was not until 1895 that

---

[215]Saari [1990] gives a clear account of the derivation of this expansion. Briefly, the equations of motion for two colliding points in the collinear central force problem are given by $\dfrac{d^2 x}{dt^2} = -\dfrac{1}{x^2}$ with solution $x(t) \sim A(t - t_0)^{\frac{2}{3}}$ as $t \to t_0$, where A is a positive constant and t_0 is the time of collision, and it can be shown for the n body problem that the same rate of approach holds for collisions of any kind taking place at $t = t_0$. Assuming that $t_0 = 0$ and substituting $X(t)t^{\frac{2}{3}} = x(t)$ into the equations of motion gives

$$t^2 \frac{d^2 X}{dt^2} + \frac{4}{3}t\frac{dX}{dt} - \frac{2}{9}X = -\frac{1}{X^2}.$$

Making the change of variable $s = t^{\frac{1}{3}}$ leads to

$$s^2 \frac{d^2 X}{ds^2} + 2s\frac{dX}{ds} - 2X = -9\frac{1}{X^2},$$

which has an analytic solution in s. Then for arbitrary initial conditions, the probability of a collision between any two of the points would be infinitely small and so could properly be ignored. Weierstrass did, however, admit that he was concerned by the fact that this method did leave open the possibility that after an infinitely long period of time two points could approach each other infinitely closely without actually colliding.

Paul Painlevé formally proved that in the three body problem such a situation cannot arise.[216]

Weierstrass ended his letter to Mittag-Leffler by saying that he felt that he had provided sufficient results to validate the claim that, in general, the coordinates of the points in the three body problem could be developed in series of the form he had specified in the question. However, since he had neither provided a proof of the impossibility of triple collision nor eliminated the possibility that the mutual distances cannot become infinitely large, his claim was somewhat tenuous.

Nevertheless, since Weierstrass did consider it legitimate to suppose that, given an unlimited time interval, the coordinates in the n body problem were single-valued continuous functions of time and as such could be represented by a series as specified in the question, he did believe that a solution was possible, and so his question was then whether such a solution was actually feasible. That was why he had asked for the description of a method which would calculate successive terms of the series rather than asking for a complete expansion. In other words, he believed it was possible to give an approximate expression for the functions such that the difference between the expression and the function did not exceed a specified arbitrarily small limit within a time interval of arbitrary length. If this could be done then the function would be represented by an absolutely and uniformly convergent series, and the problem would be solved as required.

In setting the question Weierstrass had hoped to achieve a better understanding of the true nature of the motion of celestial bodies as well as obtaining a reliable result concerning the stability of the solar system. He had little doubt that the latter could be achieved, even without a solution which was valid for infinite time.

He explained that earlier attempts to obtain a solution had resulted in the coordinates of the planets or variable orbital elements being represented by series of the form

$$\sum_{\nu_1,\nu_2,\ldots} \{C_{\nu_1\nu_2}\ldots\sin(c_0 + \nu_1 c_1 + \nu_2 c_2 + \ldots)t\},$$

where ν_i are integers, t is the time and $(C_{\nu_1\nu_2\ldots}, c_1, c_2, \ldots)$ are independent of t. Since it had been shown that, under certain assumptions, such series do formally satisfy the differential equations, what remained to be resolved was whether such series were convergent and thus true expressions of the quantities to be represented. Because this problem was clearly one of the fundamental issues raised by the competition question, its treatment provided Weierstrass with a criteria on which to base his judgement.

It was unfortunate that Weierstrass never completed his report, but it seems very probable that his analysis of [P1] would have been largely based on the letters described above. It is clear from the later correspondence that his delay in producing a report was due to his continuing difficulties over parts of the memoir that he felt it necessary to master and that he considered insufficiently explained.

---

[216] Painlevé's contribution and the question of noncollision singularities in the n body problem are discussed in Chapter 8.

He did however make one further comment concerning [P1]; that was to criticise Mittag-Leffler for the letter he had sent to the Secretary of the Paris Academy announcing the results of the competition.[217] It seems that the total French triumph had proved rather hard for the German mathematicians to bear and, in particular, exception had been taken to Mittag-Leffler's description of Poincaré's memoir as being one of "... *the most important pieces of mathematics of the century ...*".

While the Germans may well have been justified in detecting an element of self-interest in Mittag-Leffler's remark, it is fair to argue that history has proved him right.

6.3. Gyldén

One of the first people outside the commission to hear about Poincaré's memoir was Gyldén. As a member of the editorial board of *Acta*, as well as being a lecturer in astronomy at the Stockholm Högskola, he was in close touch with Mittag-Leffler and well placed to hear about the result of the competition. However, unfortunately for Mittag-Leffler, when Gyldén learnt of the results in Poincaré's memoir, his reaction mirrored that of Kronecker more than three years earlier. Gyldén, having seen the general report with its remarks about the discovery of asymptotic solutions, believed that he had already discovered similar results which he had had published in *Acta*.[218]

Mittag-Leffler appears to have had some idea of Gyldén's views almost immediately, because only a matter of days after the report was published he wrote to Poincaré to ask him for his opinion on a result in Gyldén's paper. The result appeared to conflict with something that Poincaré had written. The particular point at issue concerned the convergence of certain power series: Gyldén claimed that the series were definitely convergent while Poincaré had stated that the evidence for convergence was inconclusive. Poincaré responded swiftly to Mittag-Leffler but he admitted that he had found Gyldén's paper extremely hard to read.[219] In order to give a definitive answer he would have to make a much more detailed study of it, a task he was reluctant to undertake. He was therefore unable to say whether Gyldén's method led to a proof of either convergence or divergence, although he believed divergence more likely. He was also unhappy with the fact that Gyldén's method did not allow successive terms in the expansion to be deduced recurrently but involved making choices at each stage of the calculation, a feature which incorporated an element of chance into the process.

Meanwhile, Mittag-Leffler, at the King's request, had been due to give a review of Poincaré's paper at the February meeting of the Swedish Academy of Sciences. In the end illness prevented him from attending, although he had expressed reluctance to talk publicly about Poincaré's work without the support of Weierstrass's report. Gyldén, on the other hand, did attend the meeting and, moreover, did talk

---

[217] *Comptes Rendus* **108** (8) (25 February 1889), *387*.

[218] Gyldén [1887].

[219] Mittag-Leffler published Poincaré's side of the correspondence as part of the *Acta* volume dedicated to him.

Poincaré to Mittag-Leffler, 5.2.1889, No. 48, I M-L. *Acta* **38**, *163-164*.

Hugo Gyldén

about Poincaré's memoir. He declared his own position on Poincaré's results and effectively claimed priority.[220]

Once again Mittag-Leffler was placed in an awkward position. The King made it plain that he expected him to reply to Gyldén at the meeting the following month. Mittag-Leffler knew he could not rely on having Weierstrass's report in time, and so it became a matter of urgency to have detailed comments on Gyldén's paper from Poincaré.

On hearing from Mittag-Leffler about Gyldén's position Poincaré responded again and at length.[221] He made the point that the dispute brought into sharp focus the difference between mathematicians and astronomers with regard to their interpretation of convergence. He reasoned in detail against the rigour of Gyldén's method, reiterating that he believed Gyldén's method to rely heavily on questions of judgement, and, in his final letter on the subject, showed clearly why he believed that Gyldén's argument actually led to divergent series.[222]

Briefly, Poincaré began with the equation

$$\frac{d^2V}{dt^2} + n^2 sA \sin V \cos V = n^2(X),$$

where

$$(X) = \sum s_1 A_1 \sin(\lambda_1 nt + mV + h),$$

h is a constant and λ_1 and m are integers, and following Gyldén he put

$$V = V_0 + V_1, \qquad V_0 = -2\arctan e^{-\xi} + \frac{\pi}{2}, \qquad \xi = \alpha nt + c.$$

Gyldén's method then involved integrating by successive approximations and at each stage of the approximation choosing suitable values for the two constants of integration and the coefficient α.

Poincaré's argument hinged on the fact that he did not consider it legitimate for these choices to be arbitrary. With regard to the constants he believed that Gyldén's method meant that there was in fact only one particular value of the constants out of an infinite number of choices which would lead to a convergent series and hence to a proof of the existence of asymptotic solutions. Moreover, from what he could see, Gyldén's method gave no way of recognising which of the series was convergent. As far as α was concerned, he emphasised that its value was completely determined and could not, as Gyldén proposed, be changed with each new approximation, adding that αn was equivalent to what in his own memoir he had called the characteristic exponent.

Nevertheless, despite the critical appearance of his side of the correspondence, Poincaré did in fact maintain a high regard for Gyldén's work, appreciating the flexibility and practical advantages of his methods. He had not intended to demolish Gyldén but rather he had wanted to show how words such as *proof* and *convergence* take on different meanings depending on whether the user is a mathematician or an astronomer. Moreover, he was sensitive to the fact that Gyldén's approach was coloured by a practical interest in the problem which he himself did not share.

---

[220]Mittag-Leffler to Weierstrass, 22.2.1889, I M-L.

[221]Poincaré to Mittag-Leffler, 1.3.1889, No. 49, I M-L. *Acta* **38**, *164-169.*

[222]Poincaré to Mittag-Leffler, 5.3.1889, No. 50, I M-L. *Acta* **38**, *169-173.*

Hermite and Weierstrass were also drawn into the polemic. Hermite, who had first heard about the dispute from Kovalevskaya, thought Gyldén's series, like Lindstedt's, were asymptotic but carefully avoided drawing a direct comparison between the two memoirs. He had himself received a letter from Gyldén, but since it was written in Swedish he had been unable to read it, although he had deduced that it concerned the convergence question.

Meanwhile Mittag-Leffler gave his talk at the Academy and wrote in jubilation to Weierstrass and Poincaré, certain that those who had heard him had been convinced that Poincaré deserved the prize.[223] Although Gyldén had raised objections, insisting that his series were convergent for all time, he had admitted that in the neighbourhood of any set of the constants $c_1, \ldots, c_n$ there were other values for which the series did not converge.

However Mittag-Leffler's feeling of triumph was short lived. The academic community in Stockholm decided to weigh in on the side of Gyldén and, despite the fact that Poincaré's memoir was not in the public domain, adopted the view that Gyldén had indeed published proofs of everything Poincaré had done.[224] The consensus was that Mittag-Leffler's denial of Gyldén's results had been motivated by jealousy; this idea was reinforced by the mathematician Bäcklund, who drew attention to the fact that Gyldén's memoir had recently been awarded the St. Petersburg prize.

Meanwhile Gyldén himself steadfastly maintained that the values of the constants $c_1, \ldots, c_n$ for which his series diverged formed only a countable set, and so it was infinitely unlikely that the series was actually divergent. Mittag-Leffler continued to argue against him since, with Poincaré, he believed that the series were divergent not just for a countable set but for a perfect set in the neighbourhood of these constants. Moreover, he told Weierstrass that he thought Gyldén not enough of a mathematician to understand.[225]

With the publication of the memoir not scheduled for several months, the controversy gradually died down. Nevertheless when the memoir finally appeared, Gyldén did attempt to reopen the debate by writing directly to Hermite.[226] Possibly he thought he could count on Hermite's support, since Hermite was known to share his interest in the applications of elliptic function theory in celestial mechanics.[227] But Hermite was not to be drawn. He stood by the judgement of the commission, declaring his loyalty to Mittag-Leffler and Weierstrass. Shortly afterwards Gyldén sent Hermite part of his [1891] *Acta* paper for comment. This time Hermite avoided the issue completely by replying with the claim that the paper was outside his

---

[223]Mittag-Leffler to Weierstrass, 24.3.1889, and Mittag-Leffler to Poincaré, 28.3.1889, I M-L.

[224]Mittag-Leffler to Weierstrass, 15.4.1889, I M-L.

[225]Although not directly relevant to the disputes over Poincaré's memoir, it is of interest to record that in May that year Gyldén met with Kronecker in Berlin, a meeting which, within the context of competition, Mittag-Leffler would surely have viewed with some misgivings. In any case, the occasion prompted Mittag-Leffler to remark to Weierstrass that, although he had been led to believe that his two adversaries had understood each other perfectly, he suspected that Gyldén really understood as little of Kronecker as Kronecker understood of Gyldén. Mittag-Leffler to Weierstrass, 12.5.1889, I M-L.

[226]Hermite to Mittag-Leffler, 10.1.1891. *Cahiers* **6** (1985), *188-189* .

[227]Picard [1902] specifically mentions Hermite's sympathy for the work of Gyldén in this respect. See Hermite [1877].

own mathematical domain.[228] As he indicated later to Mittag-Leffler, he was not impressed by Gyldén's grasp of analysis, describing Gyldén as a ghost from a bygone age, who had been left behind as the world of analysis transformed about him.[229]

6.4. Minkowski

One of the earliest documented comments about [P2] came from the young Hermann Minkowski, a lecturer at the University of Bonn. In a letter dated 22nd December 1890 to David Hilbert, he revealed that he had studied the first third of the memoir, and what he had seen had reminded him of Dirichlet.[230]

Minkowski was also the author of the 1890 *Jahrbuch über die Fortschritte der Mathematik* report,[231] which appeared in 1893, by which time Minkowski had been promoted to an associate professor. This report, which seems to be the first mathematical commentary on [P2], was of a remarkable length. Most reports in the *Jahrbuch* merited at most a single page; Minkowski's report on [P2] ran to seven.

Since the function of the *Jahrbuch* was to provide information about the current state of mathematical research, Minkowski's priority would have been to provide a factual rather than a critical account of the memoir. Nevertheless, it is clear from the report that he had a good grasp of Poincaré's ideas. He skilfully picked out the salient features, emphasised their relative importance, and presented them in an accessible way.

There are various aspects of Minkowski's report which invite special comment: his clear and concise description of the theory of invariant integrals, in which he drew attention to the recurrence theorem; his discussion of Poincaré's use of the method of analytic continuation in the theory of periodic solutions; and the clarity with which he distinguished between Poincaré's use of the parameter μ and his use of its square root. Especially notable is the fact that he freely acknowledged the difficulties associated with Poincaré's doubly asymptotic solutions. Paradoxically, this probably indicates that Minkowski had a better understanding of the concept than most of his contemporaries, who abstained from passing comment on these solutions.

6.5. Hill

The first person to openly question some of the results in Poincaré's memoir and to do so in an entirely formal setting was Hill. On December 27, 1895, Hill

---

[228]Hermite to Mittag-Leffler, 1.3.1891. *Cahiers* **6** (1985), *193*.

[229]Hermite to Mittag-Leffler, 17.3.1891. *Cahiers* **6** (1985), *195-196*.

[230]L. Rüdenberg and H. Zassenhaus, *Hermann Minkowski Briefe an David Hilbert*, Springer-Verlag, 1973, *40*. It seems likely that Hilbert had asked Minkowski for his opinion on the memoir, since he had already told Klein that he had arranged for a report on the memoir to be made at his Königsberg seminar. See G. Frei, *Der Briefwechsel David Hilbert-Felix Klein (1886-1918)*, Vandenhoek & Ruprecht, Göttingen 1985, *72*.

[231]*Jahrbuch über die Fortschritte der Mathematik* **22**, *907-914*.

The *Jahrbuch* report was certainly not the first report of some of the ideas in [P2], although it does appear to be the first full report of the memoir. The first volume of Poincaré's *Mécanique Céleste*, which was derived from parts of [P2], was published in 1892, and a review appeared in the *Bulletin of the New York Mathematical Society* later the same year. See Chapter 7.

George William Hill

delivered the presidential address to the American Mathematical Society.[232] His speech was on the progress of celestial mechanics during the past fifty years, and, although meant for a general mathematical audience, the threads of his arguments were difficult to unravel, and in general it was hard to follow.

He began with a description of Delaunay's contribution and continued, "*Perhaps the most conspicuous labours in our subject, during the period of time we consider, are those of Professor Gyldén and M. Poincaré.*"[233] With regard to Poincaré's work, he cited both [P2] and the first two volumes of the *Méthodes Nouvelles*, which contained many of the results from [P2] reworked in an extended and clearer form. He centred his discussion primarily on the *Méthodes Nouvelles*, but most of his comments apply equally well to both works.

The fact that Hill took the opportunity presented by the occasion to query some of Poincaré's results can be partly explained by his belief that his own concerns were shared by other astronomers whom he thought would feel reassured by his criticism, especially those with less mathematical insight than himself. He may also have thought that a straightforward presentation of the results in a form accessible to astronomers would be somewhat superfluous, as Poincaré himself had already published a simplified account.[234] Another one might have been considered at best repetitious or at worst confusing.

Hill was particularly concerned by Poincaré's proof of the divergence of Lindstedt's series, a result which was of great practical importance to astronomers. Prior to the speech he had written an article in direct response to Poincaré's proof which had focused on the case where the mean motions are incommensurable.[235] In the article Hill demonstrated the existence of a class of cases where convergence can be shown, although he made no attempt to disprove Poincaré's argument. In the speech he aimed at reinforcing the article, and although in both cases he quoted results from [MN II] rather than [P2], it was the essential principle of the divergence of the series which was at issue. He questioned Poincaré's assertion that the convergence of Lindstedt's series would imply the nonexistence of asymptotic solutions, arguing that this was an irrelevant observation since the domains of the two things were quite distinct, i.e., where Lindstedt's series were applicable there were no asymptotic solutions and vice versa.

Since these objections concerned what Poincaré considered to be one of his most important results, and since their author was someone whose academic integrity Poincaré respected, he responded immediately.

In [1896], Poincaré's reply to Hill's first article appeared in the *Comptes Rendus* for March 2. He made it clear that there was no contradiction between their results—they had both, and in a similar way, proved the existence of cases where the series converge—but he did emphasise that it was possible for the convergence not to be uniform.

In [1896a] Poincaré countered most of the claims made in Hill's speech. Hill had believed that the series converged provided the variables remained within a

---

[232]Hill [1896a].

[233]Hill's view of Gyldén's work is described in Chapter 2.

[234]Poincaré [1891].

[235]Hill [1896] and Poincaré [MN II, *277-280*].

certain domain. Poincaré showed that the series could not converge in any part of a domain which contained a periodic solution and that every domain, however small, contained a periodic solution. Thus if the series were convergent, they could only be convergent for certain discrete values of the variables and could not be convergent for values between given limits, no matter how small the limits.

Hill had also drawn attention to the asymptotic solutions and the role of the associated characteristic exponents. His objection concerned the actual use of asymptotic solutions. He reasoned that since most of practical astronomy is concerned with systems which describe almost circular motion, a first approximation can be given by a periodic solution. By assuming this was the case, he was led to the situation where all the characteristic exponents were imaginary, and thus the coefficients of stability were real and negative, which was a situation of no interest to the working astronomer.

Poincaré pointed out that Hill's premise was based on the mistaken idea that asymptotic solutions could only exist when the variables satisfied certain inequalities. He made it clear that he had actually proved the existence of asymptotic solutions for the restricted three body problem in any domain, however small, for sufficiently small values of the perturbing mass. He attributed Hill's error to the fact that he had only considered periodic solutions of the first kind.

6.6. Whittaker

In 1898 Edmund Whittaker, then a fellow at Trinity College, Cambridge, was asked by the British Association for the Advancement of Science to draw up a report on the current state of planetary theory.[236] He responded with a substantial review of recent work on the three body problem [1899]. Whittaker's report, which was essentially an exhaustive account of the development of dynamical astronomy from 1868 to 1898 (the dates being chosen to coincide with the publication of the last volume of Delaunay's *Lunar Theory* and the third and last volume of Poincaré's *Méthodes Nouvelles*), naturally included a detailed account of [P2], which was the first commentary on the memoir to be published in English.

---

[236]Whittaker became one of the most influential British mathematicians of his generation. He was appointed Astronomer Royal for Ireland in 1906, and was elected to the chair of mathematics at Edinburgh in 1912, a post which he held until his retirement in 1946. His great treatise *Modern Analysis* [1927], which was first published in 1902, was followed only two years later by the first edition of his comprehensive *Analytical Dynamics* [1937]. The latter, which remains a standard work on the subject, includes a thorough introduction to the three body problem and contains much of his own research related to topics in dynamical astronomy.

Several of Whittaker's early papers are of interest in relation to the work of Poincaré. Of note are [1901], in which he gave a new method for expressing the solution of a dynamical problem in terms of trigonometric series; [1902] and [1902a], in which he established a new criterion for finding periodic solutions of the differential equations of dynamics and for the restricted three body problem; and in particular [1917], which concerns his discovery of the adelphic ("brotherly") integral of a dynamical system. The adelphic integral is associated with infinitesimal contact transformations between adjacent periodic solutions, and Whittaker showed that when such an integral can be constructed, the equations can be integrated and the convergence difficulties indicated by Poincaré [P2, *470*] overcome. Whittaker also gained international recognition through an article on perturbation theory and orbits which he contributed to Klein's *Encyklopädie* [1912]. For a sensitive and informative biography of Whittaker, see McCrea [1957].

As befitted the nature of the report in which it appeared, Whittaker's review of [P2] was an objective summary rather than a subjective discussion. Nevertheless, echoing Minkowski's treatment in the *Jahrbuch*, Whittaker afforded the memoir greater attention than any of the other works included in his review. In contrast to Hill, his treatment of [P2] was both complimentary and easy to follow. He began: "*A new impetus was given to Dynamical Astronomy in 1890 by the publication of a memoir by Poincaré.*"[237] He then gave a clear and concise description of many of the ideas discussed in the memoir: invariant integrals, stability, periodic solutions, characteristic exponents, asymptotic solutions, doubly asymptotic solutions, and periodic solutions of the second class. He explained Poincaré's terminology and emphasised important results, such as the recurrence theorem and the theorem concerning the nonexistence of any new single-valued integrals. His concluding remark about the final section of the memoir was, however, somewhat ambiguous in that he did not make it clear that Poincaré was raising questions concerning the general n body problem rather than solving them.

Curiously, given the extent of his report, Whittaker made no attempt to relate [P2] to Poincaré's earlier papers on differential equations beyond a single reference to his result concerning the conditions for stability. Nor did he attempt to describe Poincaré's geometric representation and the innovative technique of using a transverse section to make the problem more tractable. This may have been because he felt that the conceptual difficulty of the ideas would distract from the actual results, although that had not earlier prevented him from revealing some of the complicated details of Gyldén's method.

With regard to the doubly asymptotic solutions, he simply described them as being "*...approximately periodic when $t = -\infty$ and $t = +\infty$, but not periodic in the meantime.*"[238] While this is certainly true, it hardly gave an indication of the complex nature of the behaviour of these solutions. Admittedly Poincaré himself had not stressed this point in [P2], but he certainly did in [MN III]. But Whittaker did not mention the complexity aspect in his review of [MN III] either. It is possible that he thought it inappropriate to emphasise these solutions, since the probability of their appearance in reality was negligible. This seems unlikely, however, since the same is also true of all Poincaré's periodic solutions. Nevertheless, since he again passed over the point in his treatise on analytical dynamics, perhaps it was because he felt his own understanding was not sufficiently adequate to provide a discussion.

6.7. Other commentators

The continuing interest in the three body problem in the decade following Whittaker's report was described by Edgar Lovett [1912], who charted mathematical developments relating to the n body problem between 1898 and 1908. Apart from the burgeoning literature on the problem by way of journal articles, the period was especially notable for the publication of Hill's *Collected Works*, Moulton's

---

[237]Whittaker [1899, *144*].
[238]Whittaker [1899, *149*].

Introduction to Celestial Mechanics and Whittaker's *Analytical Dynamics*. Furthermore, by the end of the decade the publication of Poincaré's lectures on celestial mechanics [LMC] was almost complete, and the publication of the *Collected Papers* of the applied mathematician George Darwin, the pioneer of the quantitative study of periodic orbits, was about to begin.

Significantly, part of the structure of Lovett's essay leads straight back to Poincaré. Of the five headings he used, he included one on the qualitative resolution of the problem and one on periodic solutions. More specifically, he made several references to Poincaré's methods, identifying certain areas in which the influence of Poincaré's methods had been clearly felt. In particular he noted how Poincaré's preference for the canonical form of the differential equations had led to the adoption of this formulation in other investigations. Lovett also referred to developments in the theory of invariant integrals, as well as the importance of Poincaré's theorem on the nonexistence of any new single-valued integrals for the problem.

With regard to particular results in [P2], some interesting observations were made by the mathematical physicist Lord Kelvin [1891], who, having had the memoir brought to his attention by Arthur Cayley, was especially struck by the relationship between some of Poincaré's results and some conclusions of his own which he had published the previous year. In particular, he drew attention to the similarity between Poincaré's conjecture concerning the denseness of the periodic solutions [P2, *454*] and a proposition of Maxwell's concerning the distribution of energy. Maxwell had proposed *"that the system if left to itself in its actual state of motion, will, sooner or later, pass through every phase which is consistent with the equation of energy"*,[239] which, as Kelvin pointed out, was essentially equivalent to saying that every region of space would be traversed in every direction by every trajectory. If this proposition were true, which Kelvin believed to be highly likely, then he concluded it was a necessary consequence that every motion would be infinitely close to a periodic motion. In addition, he also commented on the agreement between Poincaré's results and his own results on the instability of periodic motion, observing that *"Poincaré's investigation and mine are as different as two investigations of the same subject could well be, and it is very satisfactory to find perfect agreement in conclusions."*[240]

As Brush [1966] and Gray [1992] have described, one of the first of Poincaré's ideas from [P2] to emerge in a different context was that of his recurrence theorem. This was because the theorem appeared to demonstrate the futility of contemporary efforts to deduce the second law of thermodynamics from classical mechanics. In 1896 a debate took place in *Annalen der Physik* between Ernst Zermelo, who believed that Poincaré's theorem disproved the absolute validity of the second law of thermodynamics, and Ludwig Boltzmann, who believed in the correctness of Poincaré's theorem but disputed Zermelo's application of it.[241] According to Zermelo, Poincaré's theorem implied that there were no "irreversible" processes at work, and hence the concept of a system with continuously increasing entropy was invalid. Boltzmann's defence was that the theorem was evidence of sudden brief

---

[239]Quoted in Thomson [1891, *512*].
[240]Thomson [1891, *512*].
[241]Translations of the papers by Zermelo and Boltzmann are contained in Brush [1966].

moments of decreasing entropy but that the statistical nature of his kinetic theory predicted that these moments would be so far apart that they would never actually be observed, and so entropy would in general increase. Although Zermelo and Boltzmann's personal debate came to an end within a year, the controversy continued to arouse interest and eventually became one of the sources for the foundation of modern ergodic theory.[242]

Further attention was drawn to Poincaré's work on the three body problem by his compatriot and predecessor in the chair of celestial mechanics at the Sorbonne, Félix Tisserand. In the fourth and final volume of his acclaimed *Mécanique Céleste* [1896], which was published in the year of his death, Tisserand included a chapter which consisted of Poincaré's own summary [1891] of [P2], together with some further explanations about Poincaré's periodic solutions.

Various aspects of [P2] and its underlying role in the *Méthodes Nouvelles* were naturally mentioned in Poincaré's numerous obituaries.[243] In addition, Volume 38 of *Acta*, which was dedicated to Poincaré, included two long articles describing his work: one on his mathematics by Hadamard and the other on his celestial mechanics and astronomy by von Zeipel, both articles placing a firm emphasis on the significance of the memoir.[244] Hadamard [1921] concentrated on the relationship of [P2] to Poincaré's earlier memoirs on differential equations, while von Zeipel [1921] considered its results in conjunction with the *Méthodes Nouvelles*. Of particular note is the fact that both authors quoted the passage from [MN III] where Poincaré described the complexity of the doubly asymptotic solutions. There is no doubt that the importance of these solutions had by this date been recognised, even if little further had been discovered about them. The fact that [P2] featured so strongly in these two extensive appreciations of Poincaré's career is a fitting compliment to the breadth of vision it embraced.

---

[242]See Chapter 9.

[243]See for example Baker [1914] and Darboux [1914].

[244]Mittag-Leffler had begun preparing *Acta* **38** soon after Poincaré's death, but the outbreak of the First World War meant that publication was delayed until 1921.

CHAPTER 7

Poincaré's Related Work after 1889

7.1. Introduction

After the revision of the memoir, Poincaré channelled much of his energy into amplifying the results it contained. Within two years of its publication the first volume of his celebrated *Les Méthodes Nouvelles de la Mécanique Céleste* was published. Its appearance heralded the start of an enterprise that occupied him in part for almost twenty years. The second volume was published in 1893, and the third (and final) volume was completed in 1899. The *Méthodes* was followed by its didactical counterpart, the *Leçons de Mécanique Céleste*. The *Leçons* were based on lectures Poincaré gave at the Sorbonne in his role as professor of mathematical astronomy and celestial mechanics, the position to which he had succeeded on the death of Tisserand in 1896. They contained a treatment of perturbation theory, the lunar theory, and the theory of tides, and were published in three volumes between 1905 and 1910, although according to Sarton [1913, *10*] the project was unfinished.

As well as producing these major works, Poincaré also published several more papers on different topics in celestial mechanics, some of which were connected to ideas which had appeared in [P2] and the *Méthodes*, and some of which were in response to the work of other mathematicians. There were, for example, several papers on the expansion of the perturbation function, two notes connecting the principle of least action with the theory of periodic solutions, and papers on the form of the equations in the three body problem. Other related papers included a correction to Bruns' theorem on the integrals of the three body problem, discussions of Gyldén's *horistic* methods, and some general articles.

There were also two important papers in which Poincaré continued his research into the periodic solutions of the three body problem, but outside the specific context of celestial mechanics. The first of these, which he originally presented to the American Mathematical Society at the St Louis Congress in 1904, was an investigation into the geodesics on a convex surface. This paper centred on the closed geodesics, since they enjoy an analogous role to the periodic solutions in the three body problem. The second was the paper in which he announced what is today known as his *Last Geometric Theorem*. This paper came to prominence not only because of the importance of the theorem it contained, but also because, despite strenuous efforts, Poincaré had been unable to provide a general proof.

7.2. "Les Méthodes Nouvelles de la Mécanique Céleste"

On the occasion of presenting the medal of the Royal Astronomical Society to Poincaré in 1900, George Darwin, in describing *Les Méthodes Nouvelles de la*

Mécanique Céleste,[245] said, "*It is probable that for half a century to come it will be the mine from which humbler investigators will excavate their materials*".[246] Darwin was, however, somewhat conservative in his outlook. Had he omitted the word "half" his prediction would still have been fulfilled. Since its publication almost a hundred years ago, Poincaré's *Méthodes Nouvelles* has continued to attract and delight mathematicians, providing a rich and varied source for researchers in celestial mechanics and dynamical systems.[247] Moreover, it is largely through the *Méthodes Nouvelles* that the contents of [P2] have become so widely known, for it contains the memoir's principal ideas in a more fully explained and developed form. A greater number of applications of the theory are included, as well as a substantial amount of new material, with the focus of attention being as much on the general three body problem as on the restricted problem.

Volume I, which was published in early 1892, essentially covered the analytical part of the theory. Of the topics discussed in [P2], it contains a fuller treatment of periodic solutions,[248] characteristic exponents, asymptotic solutions and the non-existence of new single-valued integrals. In addition, there is a long chapter on the expansion of the perturbation function.

The second volume, which appeared in the following year, was devoted to the methods of contemporary dynamical astronomers, namely Newcomb, Gyldén, Lindstedt and Bohlin. Most of the material was completely new, although there is an overlap with [P2] in the discussion of Lindstedt's series, and the reference to Bohlin's series which Poincaré made in the *Introduction* to [P2] is clarified.

The final volume, which is characterised by Poincaré's geometrical ideas, was published in 1899. Here Poincaré returned to the subjects of invariant integrals, stability, and periodic solutions of the second class and doubly asymptotic solutions, and also included a discussion of what he now called periodic solutions of the second species, the existence of which he had conjectured at the end of [P2].

7.2.1. Volume I. In the opening chapter of Volume I Poincaré provided a fuller introduction to both the general and the restricted three body problems than he had done in [P2]. This included placing a greater emphasis on the role of the Hamiltonian form of the equations

$$(40) \qquad \frac{dx_i}{dt} = \frac{\partial F}{\partial y_i}, \qquad \frac{dy_i}{dt} = -\frac{\partial F}{\partial x_i},$$

$$F = F_0 + \mu F_1 + \mu^2 F_2 + \cdots$$

by giving an outline of Hamilton-Jacobi theory and showing how the number of independent variables is reduced through the use of the classical integrals.

He enhanced his earlier treatment of periodic solutions by including several new applications of the theory, and his conviction of the importance of these solutions

---

[245] An English translation of *Les Méthodes Nouvelles de la Mécanique Céleste*, edited with a comprehensive introduction by D. L. Goroff, was published in 1993. See Goroff [1993].

[246] Darwin [1900, *412*].

[247] See the Foreword by J. Kovalevsky to the Blanchard edition of [MN I], 1987.

[248] Poincaré's theory of periodic solutions in [MN I] was discussed by Picard [1896, Chapter VIII]. For a modern commentary on Picard's account see Mawhin [1994].

is uncompromising. His description of them as "... *the only breach by which we can penetrate a fortress hitherto considered inaccessible*"[249] is well known.

Furthermore, the conjecture concerning the denseness of the periodic solutions is more strongly affirmed. He now proposed that given any particular solution of equations (40) it should be possible to find a periodic solution (which may have an extremely long period) such that the difference between these two solutions is as small as desired for any given length of time.[250]

The classification of the three different kinds of periodic solutions first described in [1884a] is reintroduced, and the conditions under which the solutions exist are carefully described. The first kind, in which the inclinations are zero and the eccentricities very small, are the analytic continuation of the solutions of the circular two body problem; the second kind, in which the inclinations are zero and the eccentricities finite, are generated from the solutions of the elliptic two body problem; and the third kind, in which the inclinations are finite and the eccentricities very small, are generated from an elliptic solution not in the same plane as the motion of the primaries. In other words, all Poincaré's periodic solutions are solutions which are the analytic continuation of solutions of the two body problem and thus are only valid for small values of the mass parameter.

In [P2] Poincaré had proved the existence of periodic solutions in the restricted problem, which was equivalent to proving the existence of periodic solutions of the first kind. However, his proof depended on the nonvanishing of a particular Hessian, a condition which did not hold for the general three body problem. To establish the existence of periodic solutions of the second and third kinds he first had to establish the conditions under which periodic solutions would exist when the Hessian was equal to zero.[251]

One issue that Poincaré had not raised in [P2] was the practical use of his periodic solutions. Since his analysis had shown that the probability of such solutions occurring was essentially negligible, it was not immediately obvious what practical purpose they could serve. He now made it clear that their value was not in the solutions per se but in the fact that they could be used as starting points for approximating other solutions. In the case of solutions of the first kind, which included Hill's solution, he identified as particular examples both Laplace's theory of the satellites of Jupiter and Tisserand's study of the motion of Hyperion, a satellite of Saturn.[252]

With regard to the theory of asymptotic solutions, Poincaré followed the same reasoning as in [P2] but provided a more complete theory, expanding on several points that he had previously left unexplained.

---

[249] Poincaré [MN I, *82*].

[250] Schwarzschild [1898], in his endeavour to prove the conjecture, gave a phase space interpretation in which he proposed that arbitrarily close to any point in phase space there is a point representing a periodic solution. A rigorous proof was eventually provided by Hopf [1930].

[251] Szebehely [1967, *437*] notes that Poincaré's proof of the existence of periodic solutions of the second kind contains an error which was pointed out by Wintner [1931]. See also Sternberg [1969, II, *275*] and Siegel and Moser [1971, *182*].

[252] Although Poincaré gave no practical illustrations of periodic solutions of the second and third kinds, these are also found in nature as, for example, in the motion of comets.

In order to accommodate the general three body problem in his discussion of the nonexistence of any new single-valued integral, Poincaré significantly reworked his earlier argument. As in [P2], he began by assuming the existence of another independent integral Φ, but this time he introduced Poisson brackets. As he observed, the existence of Φ implies the condition $[F, \Phi] = 0$. This Poisson bracket can then be expanded to give

$$[F_0, \Phi_0] + \mu([F_1, \Phi_0] + [F_0, \Phi_1]) + \cdots + \cdots = 0,$$

and, if this expression is true, then each of the Poisson brackets is equal to zero. To prove the invalidity of this expression, he first proved that Φ_0 is not a function of F_0 and that it is independent of y. Thus since F_0 is also independent of y,

$$-\sum \frac{\partial \Phi_0}{\partial x_i} \frac{\partial F_1}{\partial y_i} + \sum \frac{\partial F_0}{\partial x_i} \frac{\partial \Phi_1}{\partial y_i} = 0,$$

and since F_1 and Φ_1 are both periodic with respect to y they can be expanded as exponentials in the form $e^{[\sqrt{-1}(m_1 y_1 + \cdots + m_n y_n)]}$, where m_i are positive or negative integers. Considering F_1 and Φ_1 in their exponential form leads to the equation

$$B \sum m_i \frac{\partial \Phi_0}{\partial x_i} = C \sum m_i \frac{\partial F_0}{\partial x_i},$$

where B and C depend only on x. This equation shows that the Jacobian of F_0 and of Φ_0 with respect to any two x_i must vanish, providing B does not vanish, and that this occurs for all values of x such that $\dfrac{\partial F_0}{\partial x_i}$ are commensurable. Thus in any domain, however small, there is an infinite system of values of x for which the Jacobian vanishes, and since the Jacobian is a continuous function, it must vanish identically. But the vanishing of the Jacobian implies that Φ_0 is a function of F_0, which is contrary to the original assumption, and so the equations cannot admit any other single-valued integral. To complete the proof Poincaré also dealt with the cases where one or more of the coefficients B vanish and where the function F_0 does not depend on all the variables x_i.

Finally, he applied these results to the restricted problem, the planar problem and the general problem. In the latter two cases, he found necessary but not sufficient conditions for the existence of another integral of the equations. He proved that these conditions, which were in the form of relations between the coefficients in the expansion of the perturbation function F_1, did not exist. From this he could conclude that there were no new transcendental or algebraic integrals for the three body problem, but only providing μ was sufficiently small.

In the case of the general problem, proving that these conditions did not exist required a detailed analysis. In order to complete it Poincaré was led into a discussion of the perturbation function. It was not a topic that had arisen in [P2], although he had previously drawn attention to it in a note in the *Comptes Rendus*

[1891b]. Briefly, when the mean motions are incommensurable, then, due to the presence of small divisors, certain terms in the perturbation function, independent of their order, may acquire relative importance. It is not usually necessary to calculate these terms exactly, since what is important is to recognise whether they are negligible or not, and for this purpose an approximate value will suffice. Poincaré therefore looked for an approximate expansion of the function, and he became concerned with evaluating what he defined as the principal part of the function. He found he could do this by using Darboux's method for finding the coefficients of high-order terms in a Fourier or Taylor series that can be applied when the analytic properties of the functions represented by the series are known, although he first had to extend the method to functions of two variables from functions of a single variable.

As Poincaré himself observed, there was a sense in which his result concerning the nonexistence of integrals was more general than the one given earlier by Bruns [1887]. Bruns had simply proved that the ten classical integrals were the only independent algebraic integrals of the three body problem, whereas Poincaré had proved not only that there was no new transcendental integral valid for all values of the variables but also that no such integral existed even when confined to a restricted domain. But, conversely, Poincaré observed that there was another sense in which Bruns' result was more general than his: Bruns' proof was valid for any value of the masses, whereas Poincaré's method would only apply providing the masses were sufficiently small.[253]

7.2.2. Volume II.

Throughout his researches Poincaré had become increasingly aware of the differences that had evolved between the perceptions of mathematicians and of astronomers as to what constituted a solution to a problem in celestial mechanics. The difficulty was that these differences often led to what appeared to be inconsistent results, as exemplified in his controversy with Gyldén, which had involved their different interpretations of the concept of convergence.

In an effort to clarify the situation and to explain the apparent discrepancies that had arisen, Poincaré made a detailed analysis of several of the astronomers' methods for solving the equations of motion. His aim was to produce an account of the methods which would facilitate comparison between them rather than one which would make them amenable to numerical calculation.

Each of the methods Poincaré had chosen to discuss represented an attempt to expand the coordinates in the planetary theory as series in which all the terms were periodic, the secular terms having been eliminated. One issue in which he had a special interest was the convergence of the series derived.

---

[253]Painlevé [1897a, 1898] gave an extension to Bruns' theorem which showed that, apart from the classical integrals, the n body problem has no other integrals which are algebraic functions of the velocities. In [1900] he made the corresponding extension to Poincaré's theorem. Bruns' and Poincaré's theorems were more restrictive in that they only allowed for integrals which are functions of both the coordinates and the velocities.

Cherry [1924] proved that Poincaré's theorem no longer holds if the restriction that the integral must be expanded in powers of μ is relaxed.

Wintner pointed out [1941, *97, 241*] that since Bruns' and Poincaré's results are only valid for unspecified values of the masses, they are void of actual dynamical interpretation.

To illustrate how mathematicians and astronomers differed over this question, Poincaré compared the possible interpretations of the two series[254]

$$\sum \frac{(1000)^n}{n!} \quad \text{and} \quad \sum \frac{n!}{(1000)^n}.$$

He argued that a mathematician would consider the first convergent and the second divergent, while an astronomer would label them the other way round. However, he was not making a judgement about right and wrong. He considered both approaches valid in their respective domains—the first in theoretical research and the second in numerical application—his point being that it was essential to know which domain was under consideration before a decision was taken on which approach to adopt.

As Poincaré had made clear in his introduction, from a practical point of view, the question of whether the series were actually convergent or not was not the important issue. An asymptotic series, although divergent, can provide a very good approximation to a function and can be of great practical value. What is important is to have an idea of the upper limit of the error involved in using such series and to appreciate that these series cannot be used to establish theoretical results such as the stability of the solar system. Poincaré's objective in making this distinction was not in any way aimed at devaluing the work of astronomers, but rather he wanted to resolve possible misunderstandings. Indeed, he emphasised the legitimacy of asymptotic expansions in practical work, quoting results from his earlier paper on the topic [1886a].

Poincaré began with a discussion of Lindstedt's method,[255] acknowledging its equivalence with that put forward by Newcomb [1874]. In [1883] Lindstedt had used successive approximations to integrate a second-order differential equation of the form

$$\frac{d^2x}{dt^2} + n^2x = \mu\Phi(x,t),$$

and he had endeavoured to show that the solution was of the form of a wholly trigonometric series. Using Hamilton-Jacobi theory Poincaré generalised the method by developing a full canonical analogue which included results from his own paper [1889].

His problem was to calculate the coefficients of the series

$$x_i = x_i^0 + \mu x_i^1 + \mu^2 x_i^2 + \cdots \qquad (i = 1, \ldots, n)$$
$$y_i = y_i^0 + \mu y_i^1 + \mu^2 y_i^2 + \cdots$$

which formally satisfy the canonical equations. This turned out to be fairly straightforward, although due to the appearance of certain terms in the denominators of the coefficients, he had to make the assumption that the frequencies were incommensurable, which once again introduced the problem of small divisors. He also included a refined version of his [1889] proof of the redundancy of Lindstedt's symmetry condition. This was the question he had originally resolved in [1886b] through an application of Green's function but had completely reworked in [1889] by using the Hamilton–Jacobi equation in connection with a generating function.

---

[254]See also Chapter 2, section 2.3.5.

[255]As a result of Poincaré's treatment of Lindstedt's method in the *Méthodes Nouvelles*, the method is now a well-established perturbation method in applied mathematics. See Arnold [1988, *175-179*].

In addition, he also identified and resolved two particular difficulties with the method in connection with the general three body problem. The first, which he had described in the *Comptes Rendus* [1892], concerned the condition that for the method to be valid there should be no linear relationship between the mean motions. But in the general three body problem the mean motions are not only those of the two planets but also those of the perihelions and nodes; and in the first approximation, that is, in Keplerian motion, the perihelion and node are fixed, and hence the mean motions are zero. He overcame this problem by making a suitable change of variable. The second problem was related to the magnitude of the eccentricities, the squares of which enter into the denominators of the expansions and cause problems when very small. Poincaré countered this difficulty by taking as a starting point a periodic solution rather than a Keplerian ellipse.

With regard to the divergence of Lindstedt's series, Poincaré, perhaps with Weierstrass's comments in mind, went into the question in a more detailed way than he had in [P2]. In his discussion of the method itself he had shown that the Hamiltonian equations (40) could be satisfied by series of the form

$$(41) \qquad x_i = x_i^0 + \mu x_i^1 + \mu^2 x_i^2 + \cdots \qquad (i = 1, \ldots, n)$$
$$y_i = w_i + \mu y_i^1 + \mu^2 y_i^2 + \cdots ,$$

where the coefficients x_i^k (or y_i^k) were periodic functions of $w_i = n_i t + \varpi_i$, and the frequencies were given by the series

$$n_i = n_i^0 + \mu n_i^1 + \mu^2 n_i^2 + \cdots ,$$

and, in the case of two degrees of freedom, the coefficients x_i^2 could be represented by series of the form

$$(42) \qquad x_i^k = A_0 + \sum B_{m_1 m_2} \frac{\sin(m_1 w_1 + m_2 w_2) + h_{m_1 m_2}}{m_1 n_1^0 + m_2 n_2^0}.$$

To prove that the series did provide a valid solution to the differential equations, it was a question of proving the convergence of both the series (41) and (42). If the series (42) were uniformly convergent, then—as Poincaré had shown in [1884b]—the absolute value of the coefficients

$$(43) \qquad \frac{B_{m_1 m_2}}{m_1 n_1^0 + m_2 n_2^0}$$

had to be bounded, in which case n_1^0 and n_2^0 had to be incommensurable unless of course $B = 0$.

Nevertheless, even if the frequencies are incommensurable the series can still be divergent, since it is always possible to find a B such that the denominator is as small as desired and the absolute value of the coefficients (43) is therefore unbounded. But, on the other hand, it is also possible to choose the values of n_1^0 and n_2^0 so that the series are convergent—Poincaré's example was the case where the ratio of the frequencies is incommensurable but the square is commensurable. Furthermore, since the frequencies are determined by observation, they can only be given to within a certain approximation, and therefore it is always possible to choose them so that the series are convergent but at the same time remain within the limits of the approximation. The next question is whether the series are convergent for all values of the constants of integration x_i^0 within a certain interval (since the n_i^0 depend on the x_i^0). It turns out that although in general this is not the case, in practice it is always possible to restrict sufficiently the calculation so that the series (42) are only composed of a finite number of terms and the series (41) can be formed.

It therefore remains to discuss the convergence of series (41). In this case, there are two questions to consider. First, are the series uniformly convergent for all values of μ and x_i^0 within a certain interval? And second, are the series uniformly convergent for all sufficiently small values of μ for suitably chosen values of x_i^0? Poincaré easily showed that the answer to the first question was "no": the series were in general divergent. With regard to the second he made a distinction between whether or not the frequencies were dependent on the parameter μ. In the case where the frequencies depended on the parameter, he observed that for sufficiently small values of μ it was always possible to find values of μ such that the frequencies were rationally related (since the ratio is a continuous function of μ). The series then represent a periodic solution of the Hamiltonian equations for all values of the two constants of integration ϖ_i. Hence if the series are convergent then corresponding to this ratio there are a double infinity of periodic solutions. But, as he had previously shown both in [P2] and in [MN I], this only occurs in very exceptional cases. As a result Poincaré came to the conclusion that the series (41) were not convergent, although he recognised that his argument was insufficient to establish the point with absolute rigour.[256]

In the case when the frequencies are independent of the parameter, the question is whether values of x_i^0 can be chosen so that the series are convergent, the choice of the values of x_i^0 being made by imposing some condition on the ratio of the frequencies, such as, for example, its square being rational. But in this case Poincaré was even more noncommittal. All he would say was that: "*The arguments presented in this Chapter do not allow me to affirm that this* [i.e., that the series are convergent] *cannot happen. They only allow me to say it is very unlikely.*"[257]

Thus although Poincaré had reached fundamentally the same conclusion with regard to the divergence of the series as he had in [P2], he had investigated the question much more thoroughly, and, importantly, he had cast doubt over his results in the case where the frequencies were chosen in advance.

Nevertheless, despite Poincaré's note of caution it was generally accepted that he had proved the divergence of the series and that Weierstrass must have been wrong. From what Poincaré had shown, there did not appear to be any conditions under which the series were in general convergent. But almost seventy years later it was shown that Poincaré had been right to be cautious with his conclusions, for, contrary to expectations, the question was finally resolved in Weierstrass's favour by Kolmogorov, Arnold and Moser.[258]

---

[256]Poincaré [MN II, *108*]. Moser [1973, *9*] mistakenly ascribes Poincaré's concern over the rigour of his argument to the case where the frequencies are independent of the parameter.

[257]Poincaré [MN II, *105*].

[258]An outline of Kolmogorov, Arnold and Moser's contribution is given in the Epilogue.

With regard to the work of Gyldén, Poincaré centred on the integration of the particular form of Hill's equation given by

$$\frac{d^2x}{dt^2} = x(-q^2 + q_1 \cos 2t),$$

which Tisserand [1894, Chapter I] had called the Gyldén-Lindstedt equation.

Poincaré described not only Gyldén's method of integration but also the methods of Bruns, Hill and Lindstedt. He also discussed the more difficult nonlinear equation of evection[259]

$$\frac{d^2x}{dt^2} + x(q^2 - q_1 \cos 2t) = \alpha\Phi(x,t),$$

which had been extensively analysed by Gyldén.

In the final part of the volume Poincaré considered the problem of small divisors in connection with a method devised by Bohlin. Bohlin's method was essentially an improved version of Delaunay's in that it involved the same basic ideas but without the inconvenience of numerous changes of variable. However, although the method did successfully eliminate small divisors, it had the disadvantage of generating the reciprocal problem of large multipliers. Here Poincaré makes clear the analogy between Bohlin's series and his own series for the asymptotic solutions of the restricted problem, to which he had referred in the introduction to [P2]. Moreover, he proved the divergence of Bohlin's series by using the same example as the one he had used to prove the divergence of his own series. In the final chapter Poincaré extended Bohlin's method in order to eliminate some of the difficulties which arise when the basic method is applied to the three body problem.

In this second volume Poincaré forcefully demonstrates the importance of understanding the nature of the convergence of the different series used in the expressions for the coordinates of the planets. He identified the respective advantages and disadvantages of each of the different methods for obtaining these series, while at the same time making improvements and corrections. He recognised Newcomb's and Lindstedt's methods to be the simplest, and in particular recommended them to be used when there was not a problem with small divisors. His verdict on Gyldén's methods was that they were too complex to be of any real practical help but that they were extremely valuable both in terms of the insight they lent to particular problems and in their use for overcoming specific difficulties. In the case where the mean motions give rise to the problem of small divisors, then Poincaré's analysis had shown that it was necessary to use Delaunay's or Bohlin's methods. With regard to the three body problem Poincaré expressed a preference for Bohlin's methods, which were similar to his own. But, most important of all, in contrast to what their authors had assumed, Poincaré had shown that most of these series were not convergent but were instead asymptotic expansions,[260] the methods of

---

[259]The evection, the largest lunar perturbation, is caused by the periodic variation in the eccentricity of the lunar orbit.

[260]Baker [1916] proved that the formal series solution to the equation

$$\frac{d^2x}{dt^2} + x(1 + \Psi) = 0, \qquad \Psi = A\cos\alpha t + B\cos\beta t,$$

where A and B are small, α and β are incommensurable, and A and B have a small common factor μ, was not, as Poincaré had claimed [MN II, *277*], divergent.

Newcomb, Lindstedt and Gyldén resulting in series in μ, and Bohlin's resulting in series in $\sqrt{\mu}$.

7.2.3. Volume III. The final volume of the *Méthodes Nouvelles* is the most geometric of the three. It begins with the theory of invariant integrals, which is given a much improved and more logical structure than in [P2]. The theory is applied to the general three body problem, which includes a table detailing the number of invariants in the different formulations of the problem. Also included is a long discussion on invariant integrals and asymptotic solutions. In this Poincaré proves that in the case of the Hamiltonian equations it is most unlikely that there exist any other algebraic or quadratic invariant integrals other than those already found, and he shows how this fundamental property is related to the nonexistence of any new integrals for the equations. There is also a chapter on iterative processes which contains the geometric theorems which were established at the end of the chapter on invariant integrals in [P2].

This volume also contains a discussion of stability in the three body problem in which Poincaré states three sufficient conditions:

1. The distances between the bodies can never become infinitely large.
2. The mutual distances between the bodies is never less than a given limit.
3. The system returns infinitely often arbitrarily close to its initial position.

As Poincaré observed, Hill had already proved that the first condition was satisfied in the case of the restricted problem. Poincaré did not address the second condition, but went on to establish, via his recurrence theorem—as he had done in [P2]—that the third condition is generally satisfied for the restricted problem, although the result cannot be extended to the general problem.

He made a more detailed study of periodic solutions of the second class —that is those periodic solutions which make more than one orbit around the primary—than he had in [P2]. In particular he developed the connection between these solutions and the principle of least action, a topic which he had introduced in two notes in the *Comptes Rendus* [1896c, 1897]. In the first of these notes he had shown that this principle could be used to infer the existence of different kinds of periodic solutions, where the law of attraction is some inverse power of the distance higher than the square. In the second note he used the principle to distinguish between two different types of unstable periodic solutions, showing that if the constants of motion are varied continuously it is impossible to move directly between the two types of unstable periodic solutions without passing through a stable periodic solution.

Poincaré also related his study of the periodic solutions of the second class to the periodic solutions discovered by Darwin through numerical analysis and found that almost all of Darwin's results were in accordance with his own theory. The one discrepancy concerned the stability of a certain family of Darwin's solutions, where Poincaré identified an error in Darwin's theory.[261]

At the end of [P2] Poincaré had postulated the existence of a system in which two very small bodies describe orbits around one large body, and these orbits are

---

[261]The error is explained in Chapter 8 in the discussion on Darwin's work.

such that collisions are narrowly avoided at definite intervals. If these orbits did exist they would be ellipses with elements which would remain almost constant except near each "collision" point, where they would suddenly change dramatically. In other words, Poincaré had considered the possibility that all the elements in the system could vary in such a way that the resulting motion was periodic. He now investigated this idea further.

He considered equations (40) with p degrees of freedom, with periodic solutions of period T such that when t is increased by T, the variables $y_1, \ldots, y_p$ increase by $2k_1\pi, \ldots, 2k_p\pi$ where $k_1, \ldots, k_p$ are any integers. In the case of the three body problem $y_1, \ldots, y_6$ represent the mean longitudes, the perihelions and the nodes of the planets, and F_0 depends only on the variables x_1 and x_2 which are proportional to the square roots of the major axes. A solution will then be periodic if the differences between the y increase by multiples of 2π as t increases by a period T, and in this case F only depends on these differences. If $2k_1\pi, \ldots, 2k_5\pi$ are the increases in

$$y_1 - y_6, \qquad y_2 - y_6, \qquad y_3 - y_6, \qquad y_4 - y_6, \qquad y_5 - y_6,$$

as t increases by a period, then, as Poincaré had previously established in Volume I, there are periodic solutions for arbitrary values of k_1, k_2 providing k_3, k_4 and k_5 are zero. Poincaré now considered the idea of solutions in which the five integers k take arbitrary values.

From considerations of continuity and the fact that very little modification is required to the function F in order to regain the original Hamiltonian equations, it appeared likely that such solutions did exist. Although when the mass parameter μ is zero the two planets follow Keplerian orbits, which indicates that k_3, k_4 and k_5 must be zero. To counter this difficulty Poincaré assumed that the two planets describe almost Keplerian orbits except at a certain moment when their mutual distance becomes small enough to produce a strong perturbation, as a result of which their perihelia and nodes change by large amounts. For such orbits the perihelia and nodes are certainly not fixed and so for $\mu = 0$, k_3, k_4 and k_5 cannot be zero. Although Poincaré claimed that his procedure was sufficient to prove the existence of such orbits, which he called periodic solutions of the second species, it is now not clear that his claim is justified.[262]

The final chapter of Volume III was devoted to a discussion of doubly asymptotic solutions. This contained essentially the same analysis as [P2] but with one important addition. In [P2] Poincaré had shown that corresponding to each unstable periodic solution there was a system of asymptotic solutions, each set of which formed an asymptotic surface whose intersection with a transverse section formed an asymptotic curve. He had distinguished two families of asymptotic solutions, one which approached the periodic solution as $t \to -\infty$ and one which approached it as $t \to +\infty$. He had then proved that two asymptotic curves can only intersect if they come from different families, calling such an intersection a doubly asymptotic solution. However, he had only considered the possibility of doubly asymptotic solutions arising from different families of asymptotic solutions associated with the same unstable periodic solution. These are what he called *homoclinic* solutions.

---

[262]Levy [1912] questions the sufficiency of Poincaré's proof and observes that P. Semirot, in his thesis published in 1943, gives some counterexamples.

Now he proposed the idea that a different type of doubly asymptotic solution could arise from asymptotic solutions associated with two different unstable periodic solutions and called these *heteroclinic* solutions.

He established the existence of homoclinic solutions in the restricted three body problem, as he had done in [P2], but this time adding an unequivocal statement about their bewildering complexity:

> *When one tries to depict the figure formed by these two curves and their infinity of intersections, each corresponding to a doubly asymptotic solution, these intersections form a kind of net, web, or infinitely tight mesh; neither of the two curves can ever intersect itself, but must fold back on itself in a very complex way in order to intersect all the links of the mesh infinitely often.*

> *One is struck by the complexity of this figure that I am not even attempting to draw. Nothing can give us a better idea of the complexity of the three-body problem and of all the problems of dynamics in general where there is no single-valued integral and Bohlin's series diverge.*[263]

Poincaré concluded with a discussion of the heteroclinic solutions in which he proved, as he had done in the homoclinic case, that the existence of one solution is sufficient to prove the existence of an infinite number.

The difficulties that Poincaré had encountered in trying to understand these doubly asymptotic solutions is evident from the fact that almost ten years had elapsed since he had first introduced the idea in [P2], and despite the time interval he had added relatively little to his original discussion, apart from the important inclusion of heteroclinic solutions. It seems likely that he did not consider the possibility of the latter in [P2] because he had discovered the homoclinic solutions as a result of realising that the asymptotic surfaces arising from one unstable periodic solution were not closed but intersecting; and it would therefore have been natural to focus only on the implications of this particular intersection. In any case he had found this result so shocking that it is not perhaps surprising that he did not consider the possibility of even more complex solutions. A homoclinic solution only involves one curve folding back on itself without self-intersecting while simultaneously cutting another curve infinitely often, but a heteroclinic solution involves two curves folding back on themselves and is correspondingly much more complicated. It is therefore small wonder to find that contemporary reviewers of the *Méthodes Nouvelles* had little to say about Poincaré's analysis of these solutions, except to reiterate his words about their complexity.

An indication of the complex behaviour of Poincaré's homoclinic and heteroclinic solutions is shown in *Figure 7.2.i*. The diagrams represent the intersections of the asymptotic surfaces with a transverse section showing the solutions in the early stages of their development. The points S and S' represent unstable periodic solutions, C and C' are two asymptotic curves corresponding to S, and D and D' are two asymptotic curves corresponding to S'. The points P and Q are then a homoclinic point and a heteroclinic point respectively.

---

[263]Poincaré [MN III, *389*].

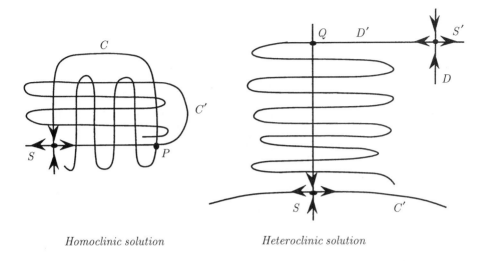

Homoclinic solution   *Heteroclinic solution*

FIGURE 7.2.i

7.2.4. Reviews. An enthusiastic review of [MN I] was given by Brown [1892] in the first volume of the *Bulletin of the New York Mathematical Society* in which he welcomed the appearance of a mathematical treatise on the problems of celestial mechanics while at the same time recognising the wider applications of Poincaré's methods. In particular, he was impressed by Poincaré's "penetrative genius" for dealing with convergence arguments and his treatment of periodic solutions with its application for the lunar theory. Brown successfully managed to communicate in a concise style unencumbered by mathematical detail the essence of most of Poincaré's ideas. However, whether due to a time constraint (the review was in print by the middle of 1892) or to his unfamiliarity with the mathematics, he glossed over the last two chapters, the one on the perturbation function and the one on the theory of asymptotic solutions, only signalling the titles and giving no indication of their contents.

Whittaker's [1899] report contained a brief review of all three volumes of the *Méthodes Nouvelles*, but due to its relationship with [P2], his observations were mostly confined to a brief outline of the new results in the second volume. As with his commentary on [P2], there is nothing in the way of subjective discussion; nevertheless, it is unexpected to find that there is no real indication of the stature of the work.

Perchot [1899], presumably prompted by the appearance of [MN III], gave an extremely detailed synopsis of [MN I] in which he noted Poincaré's emphasis on the importance of questions of convergence. Although he went into considerable mathematical detail, he essentially reproduced Poincaré's own arguments without providing any additional comments or explanations and so gave no new insights into any of the material or how it was received.

Maurice Hamy [1892, 1896, 1900], an astronomer at the Observatory in Paris, provided coherent and concise reviews of all three volumes of [MN]. Clearly meant for a general scientific audience, these reviews conveyed the spirit of Poincaré's ingenuity without getting lost in the mathematical detail.

7.3. The three body problem and celestial mechanics

Of the many other papers on the three body problem and celestial mechanics which Poincaré produced after [P2], several heralded results which later appeared in a volume of [MN] and several were refinements to results which had already been published. Included amongst these were several papers on the expansion of the perturbation function and others on the calculation and convergence of the series used to integrate the differential equations of the three body problem. These will not be discussed here, but they can be found in Volumes VII and VIII of the Poincaré *Œuvres*.

7.3.1. Bruns' theorem. In [1896b] Poincaré added further to his work on the question of the nonexistence of new integrals of the three body problem by making a correction to Bruns' [1887] theorem in which Bruns had proved the nonexistence of any new single-valued algebraic integral for the general three body problem for all values of the mass parameter.

The mistake in Bruns' paper related to an expression which Bruns had believed to be an exact differential, a condition which had to hold for his theorem to be true. Poincaré's correction involved giving a simple counterexample which demonstrated that the condition did not hold in the generality Bruns had described. Poincaré's discovery did not in fact invalidate Bruns' result, for he went on to prove that although it was possible for functions to exist which did not satisfy the exact differential condition but which satisfied all the other necessary conditions for Bruns' theorem to be true, these particular functions could not arise from the three body problem. In other words, although Bruns' argument had been technically incorrect, his conclusion regarding the integrals of the three body problem was in fact valid. Characteristically, Poincaré did not elaborate on the details of his method and only sketched his proof in the broadest outline. It was clearly a delicate analysis, for not only was an error found by Whittaker in a detailed proof supplied by Forsyth, but Whittaker's own proof, while rectifying Forsyth's error, also included an error. The latter was pointed out by MacMillan [1913] in his discussion of Poincaré's correction.

7.3.2. Gyldén's horistic methods. Poincaré's discussion of Gyldén's horistic methods can be traced through a series of notes in the *Comptes Rendus* culminating in a memoir in *Acta* [1905a]. Beginning in [1901] he raised objections to Gyldén's first horistic method, and then in [1904] turned his attention to the second method. In the latter his investigations showed that Gyldén's method could not, as Gyldén had claimed, be used to determine the general solution to the differential equations, although with certain modifications it could be used to determine a particular periodic solution. In addition he proved the falsity of Gyldén's conclusion that high-order terms in the perturbation function could never cause libration. Shortly afterwards, Bäcklund, who after Gyldén's death in 1896 had been given the responsibility for editing Gyldén's manuscripts,[264] called into question Poincaré's results, which in turn elicited a response from Poincaré [1904a]. Poincaré brought together his ideas on Gyldén's theory in a more comprehensive form in [1905a].

Although Poincaré began [1905a] by pointing out the great service Gyldén had done to celestial mechanics by the creation of new methods that had been used to great effect in, for example, the theory of small planets, he further observed that these had now been largely superseded by more convenient methods such as those of Hill and Brown, and he was unable to speak so kindly of Gyldén's later work. The paper is essentially a criticism of the method Gyldén had developed to overcome the problem of small divisors and had described in a long paper on the convergence of series used in celestial mechanics.[265]

Since the usual methods used to solve problems in celestial mechanics result in solutions with certain terms which have coefficients $\frac{b}{p^2}$ which become infinite when p vanishes, it is clearly desirable to try to improve the methods so that they do not result in terms of this type. Gyldén tried to show that if when using the traditional methods a more exact calculation is made, then these terms will not arise, and instead there will be what he called horistic terms—these being terms with coefficients of the form $\frac{b}{(\nu^2 + p^2)}$, ν being a very small but nonvanishing quantity. The point being that if Gyldén was right and these terms did exist, then they could be used to prove the convergence of the series. In addition, Gyldén had argued that the horistic method led to an important result concerning the conditions for libration. As Poincaré was to show, both these conclusions were false.

Originally Poincaré had believed that the mistakes in Gyldén's paper derived from Gyldén's misunderstanding of what was meant by a mathematical proof of convergence. Moreover, since Gyldén's method was so obscure, Poincaré felt reluctant to try to unravel the errors, especially as he thought, like Hill, that the inherent obscurity would deter anyone else from using the method, and so the errors would not get perpetuated. He had therefore been tempted to let matters rest. But on making a more careful examination of Gyldén's paper he found, apart from the problems relating to the definition of convergence, that there were also errors contained in Gyldén's first approximations that occurred at the beginning of the analysis. In addition, other astronomers and mathematicians, as for example Bäcklund, had been tempted into applying the method to practical problems and had run into difficulties. Thus Poincaré eventually felt compelled to attempt to explain the mistakes.

In addition to trying to identify the faults in Gyldén's analysis, Poincaré was also anxious to put the record straight regarding Gyldén's result about libration. In the final part of his paper, Gyldén had applied the principles of his method to the three body problem, with the result that he thought he had proved that higher order terms in the perturbation function were not responsible for libration. This deduction was based on the belief that in these higher order terms the horistic terms, which oppose libration, dominate. Poincaré, referring to his own results

---

[264]See Whittaker [1899, *144*].
[265]Gyldén [1893].

in [MN I], showed why this conclusion was false. His argument was based on the
fact—which he had already proved—that each term of the perturbation function,
however high the order, was equivalent to a system of periodic solutions. Since close
to each stable periodic solution (there being as many stable as unstable periodic
solutions) there are solutions which oscillate and cause libration, this means that
any term in the perturbation function can cause libration, providing not all the
characteristic exponents vanish, which for terms of sufficiently high order does not
occur. Thus Gyldén's conclusion is false.

Although Poincaré had been able to uncover some of the errors in Gyldén's
paper and prove the falsity of his conclusions, he was still unable to grasp fully the
intricacies of Gyldén's procedures. His final comments were somewhat reminiscent
of his remarks about Gyldén's [1887] paper discussed above:

> Several of his [Gyldén's] results are clearly correct, but they could have
> been reached by a much quicker method; a great number are clearly
> false; most of them are given in a way which is too obscure to decide
> whether they are true or false.[266]

7.3.3. General papers. Apart from the papers dealing with specific questions arising in celestial mechanics, Poincaré also wrote three papers of a more
general nature, two on the three body problem and one on the stability of the solar
system. These embraced a greater practical perspective than the other papers and
were a response to the need for a more popular exposition of his ideas.

Mention has already been made of Poincaré's synopsis of [P2] that appeared
in the *Bulletin Astronomique* [1891].[267] This was essentially an outline of the main
ideas and results from [P2], framed in such a way so as to be accessible to those,
such as astronomers, whose interest in the three body problem was motivated by
practical considerations. In this paper he concentrated on his use of the more
familiar methods of infinite series rather than on his innovative geometric insights.
He again referred to the practical value of his periodic solutions, pointing out that,
although they were clearly an artificial construction since the probability of the
initial conditions being such as to generate them was theoretically zero, they could
be used to provide a very good first approximation for the intermediate orbit.

During the same period, Poincaré also provided a second review of [P2] where
the exposition of his results regarding the restricted three body problem was almost
completely descriptive [1891a]. He illustrated the concepts using examples rather
than theoretical mathematics and even managed to avoid including a single formula,
either for a differential equation or an infinite series. Again he made a reference to
the practical value of his theoretical solutions, and, in addition, he also touched on
the relationship between some of his mathematical results and the physical question
of the stability of the solar system.

The stability of the solar system was also the subject of an article [1898] which
appeared not only in two different publications in France in 1898 but also in *Nature*
(in translation) in the same year. The article explained the basis of earlier stability
proofs, such as those supplied by Lagrange and Poisson, which were founded on

---

[266]Poincaré [1905a, *618*].

[267]This paper was also reproduced in its entirety by Tisserand [1896, Chapter 27].

methods of successive approximations. These methods showed that the variations in the elements were reduced to oscillations of a small amplitude about a mean value, and that this mean value itself was subject to oscillations. It also contained a discussion of the limitations afforded by the theoretical representation of the solar system as a system of material points subject to the exclusive action of their mutual attractions.

Acknowledging that real bodies are subject to forces other than gravitational attraction, Poincaré enquired into the nature and magnitude of what he termed these *complementary* forces. For, as he observed, if it could be shown that the effect of these forces was actually greater than the effect of the terms neglected by the approximations made in a theoretical proof of the stability, then the degree of accuracy lost through making the approximations could be legitimately ignored. The question Poincaré posed, therefore, was whether the stability was more easily destroyed by the complementary forces or by gravitational attraction.

With regard to the nature of these complementary forces, there was the recognised problem of the inconsistency of Newton's law with regard to the motion of the perihelion of Mercury. But, as Poincaré pointed out, providing any replacement law was sufficiently close to the inverse square law, it could be considered as equivalent from the stability point of view and would not affect the final outcome. However, he identified another more compelling reason that argued against stability: the second law of thermodynamics, according to which there is a continuous dissipation of the energy generated by transforming work into heat. He suggested that this manifested itself in the motion of celestial bodies, both in the continuous action of tides, which, since the bodies are not perfectly elastic, occurs even when the bodies are solid, and in the forces created by the magnetic fields of the bodies.

Having examined these forces he concluded that although the dissipation of energy resulting from their effect was extremely slow, it was still fast enough to be greater than the effect which would be imposed by those terms which were neglected by approximation in a theoretical proof of stability. In other words, from a practical point of view, Poincaré believed that the accuracy of the theoretical proof had reached its useful limit, although that did not in any way detract from the interest or value in continuing research into the purely theoretical problem.

7.4. General dynamics and "The Last Geometric Theorem"

7.4.1. Geodesics on a convex surface. As Poincaré had stressed in [P2] and [MN I], the study of periodic solutions was of the utmost importance in analysing the motion in the three body problem. However, after his initial assault on the topic, it was more than ten years before he presented another paper on the subject, and when he did it was within quite a different framework. His discovery of the complexity of the periodic and asymptotic solutions of the restricted three body problem had made him realise that to gain a greater understanding of the underlying dynamics it was necessary first to study a simpler dynamical problem than the one he had treated. The paper [1905b] he presented at the St. Louis Congress investigated just such a dynamical problem: the question of the existence of geodesics on a convex surface. On the one hand it is a problem with two degrees of freedom and so analogous to the restricted three body problem, while on the other the inherent simplification from the lack of singular points implies a constant velocity which can be regarded as given. Furthermore, it has a direct application to the three body problem, since the trajectories of the three body problem are comparable to geodesics on such a surface, the closed geodesics representing periodic solutions. Nevertheless, despite the simplification, it is still an extremely difficult problem.

The force of the paper lay in Poincaré's use of variational calculus and the method of analytic continuation. Having shown how the property of a minimum used to characterise geodesics can be used to establish the existence of closed geodesics on an ellipsoid only slightly different from a sphere, and moreover that the total number of closed geodesics must be odd, Poincaré then considered a continuous family of analytic convex surfaces depending analytically on a parameter t, connecting a sphere where $t = 0$ to a given surface where $t = 1$. He found that by continuously varying the parameter the closed geodesics appeared and disappeared in pairs. Furthermore, using topological considerations he was able to distinguish between the different types of closed geodesics by the number and arrangement of their double points. He was then able to extend his earlier result to show that on an arbitrary convex surface there is at least one closed geodesic without a double point, and, furthermore, since on an ellipsoid there are three, by continuity there are always an odd number of them. Thus Poincaré's results pointed towards the idea that there were in fact at least three closed geodesics on such a surface, although he did not attempt to prove it. Later Birkhoff established the existence of three closed geodesics on an arbitrary convex surface, subject to certain limitations [1927, *180*], and a complete solution was finally provided by Lusternik and Schnirrelmann [1930].

Poincaré also addressed the question of the stability of the closed geodesics. Here again he used his idea of characteristic exponents, and he also drew specific analogies with his results from [MN III] concerning periodic solutions and the principle of least action. Considering all the closed geodesics without a double point on an arbitrary convex surface, he found that the excess of the number of stable over the number of unstable geodesics remained constant. Furthermore, since on an ellipsoid this excess is equal to one (the largest and smallest of the principal ellipses on an ellipsoid are stable closed geodesics while the third is unstable), by continuity there is always at least one stable closed geodesic on an arbitrary convex surface.

Although extensive, Poincaré's account was by no means exhaustive. He did not, for example, consider higher dimensional ellipsoids, and with regard to his result that closed geodesics appear and disappear in pairs, he did not take into account that infinite families of closed geodesics of the same length can appear on a particular surface, as in the case of a sphere. Furthermore, although he had shown that for values of t sufficiently close to 0, it was possible to use analytic continuation to obtain from the principal ellipses on the ellipsoid an odd number of closed geodesics on an arbitrary convex surface, he gave no criteria for how far this method could be carried out.[268]

---

[268]For a discussion of the difficulties and limitations of Poincaré's paper see Morse [1934, *305-358*].

7.4.2. "The Last Geometric Theorem". Poincaré's last attack on the three body problem [1912] was also connected with the question of periodic solutions, but again the form was quite different from his original investigations. This time his arguments were based on considerations of algebraic topology. In the paper, which was published only a few weeks before his death, Poincaré announced a theorem which if shown to be true would confirm the existence of an infinite number of periodic solutions for the restricted three body problem for all values of the mass parameter μ. Furthermore, he believed that the theorem would eventually be instrumental in establishing the denseness of the periodic solutions. However, although he had been working on the theorem for two years, he had not been successful in finding a complete proof. Nevertheless, as he explained in the introduction, he felt it important to publish it despite the fact that it was in an unfinished state:

> *It seems that in these circumstances, I should refrain from any publication for as long as I am unable to resolve the question; indeed after the useless efforts that I have made for so many months, it appeared to me that it would be wisest to leave the problem to mature, while resting for several years; that would be all very well if I was certain to be able to return to it one day; but at my age I cannot be sure. On the other hand, the subject is so important (and I will search further to understand it) and the set of results already obtained so considerable, that I am resigning myself to leave them incomplete. I hope that the mathematicians who will interest themselves in this problem and who without doubt will be more successful than me, will be able to take advantage of them and use them to find the way in which they should go.*[269]

As is well known, shortly after Poincaré's death the young American mathematician George Birkhoff [1913] was indeed successful and supplied a brilliantly elegant proof, creating one of the mathematical sensations of the decade.[270]

Poincaré's theorem can be given in the following form:

THEOREM. *Suppose that a continuous one-to-one area-preserving transformation T takes the ring R, formed by the concentric circles of radii $x = a$ and $x = b$ ($a > b > 0$) into itself in such a way so as to advance the points on $x = a$ in a positive sense and the points on $x = b$ in a negative sense; then there are at least two points of the ring invariant under T.*

In fact, as Poincaré observed, to prove the theorem it is sufficient to prove the existence of just one invariant point, since topological considerations show that if there is one invariant point then there must be a second.[271]

---

[269]Poincaré [1912, *500*]. Painlevé [1912], writing on the day of Poincaré's death, described the introduction as a simple but noble testament to a life completely dedicated to the search for truth.

[270]Some measure of the effort it required on Birkhoff's part to provide the proof can be estimated from the fact that several years later he apparently admitted to losing 30 lbs. in weight while working on it. See Parikh [1991, *40*].

[271]Poincaré modestly attributed this result to Kronecker, although it essentially derives from his own index theorem for the case of the sphere [1885, *125*].

In considering the application of the theorem to the restricted three body problem, Poincaré began with the customary formulation of the problem in a rotating coordinate system with the Jacobian integral

$$J = \frac{1}{2}(x'^2 + y'^2) + H(x, y) = C,$$

where x' and y' are the components of the velocity. The motion then takes place in the plane region β defined by $H < C$, which is bounded by a closed curve α. The velocity is given in magnitude but not in direction, and at each point of α the velocity is zero. Therefore to each point of β there correspond an infinite number of *elements* (defined by a particular geodesic together with a point on that geodesic) comprising both speed and direction, and to each point of the boundary α there corresponds only one element.

Poincaré first made a topological mapping of the region β into the interior of a circle β' so that the boundary α is mapped into the circumference α'. To examine the motion in the region β', he considered a circle γ whose plane is perpendicular to the plane of β' with diameter MM', where M is a point either in β' or on α' and M' is its inverse with respect to the circle α'. Then to each element through M he made a correspondence with a point on γ, the correspondence being determined by the direction of the element through M. Thus if M is in β' there are an infinite number of related points, one for each element, and if M is on α' there is exactly one related point. Therefore each element corresponds to one and only one point in space and conversely.

The trajectories are therefore represented by the members C of a family of twisted curves, where the closed curves represent the periodic solutions, and one and only one curve passes through each point in the space. Poincaré then considered a closed curve C_0 which represents a periodic solution G_0 and an area A bounded by this curve which lies on a curved surface S. If A is simply connected and without contact, that is, no curve C other than C_0 is tangent to S at a point of A, and if P is a point of A with consequent P', then the transformation T which transforms P to P' is a point transformation of A onto itself, and as Poincaré showed, the transformation is continuous. He further observed that the transformation T admits a positive invariant integral and is therefore area preserving.[272]

If P is a point of A close to the boundary curve C_0, then the curve C_1 through P represents a trajectory G_1 close to the periodic solution G_0. Poincaré showed that when P is very close to C_0, a function which he called the *reduced argument* of P and that of its consequent P' differ by $\dfrac{2\pi}{\alpha + m}$, where $\pm i\alpha$ are the nonzero characteristic exponents for the stable periodic solution G_0, m is an integer, and the reduced argument has the property that it varies steadily from 0 to 2π around C_0.

He next considered a topological mapping of the area A onto the interior of a circle so that, using polar coordinates (x, y), C_0 becomes the circle $x = a$, and on this circle y is equal to the reduced argument. The transformation T then maps the circle into itself, and each point on the circle is advanced through the angle

---

[272]Poincaré did not prove the proposition concerning the existence of an invariant integral but instead made reference to the appropriate parts of the *Méthodes Nouvelles*.

$\dfrac{2\pi}{\alpha + m}$. By his index theorem such a mapping has an odd number of fixed points in the interior of A, each of which corresponds to a periodic solution, and at least one of which is stable. If P_0 is the fixed point corresponding to the stable periodic solution, and the coordinates are chosen such that P_0 is the centre of the circle $x = 0$, then T leaves the centre of the circle unchanged and maps the circumference onto itself in such a way that all the points are moved in the same direction through the same angle.

If C_0' represents the stable periodic solution through P_0, and P is now considered to be a point close to P_0, then Poincaré showed that in this case the increase in the value of y in passing from P to its consequent P' was $2\pi(\beta + n)$, where $\pm i\beta$ were the two nonzero characteristic exponents of the periodic solution passing through P_0 and n is a fixed integer.

Finally, he considered the iterated transformation T^p, where p is a positive integer. T^p therefore conserves both $x = a$ and $x = 0$ and is area preserving. If $T(x, y) = (X, Y)$, then the iterated transformation will give, on $x = a$,

$$Y - y = 2\pi \left(\frac{p}{\alpha + m} \right),$$

and, on $x = 0$,

$$Y - y = 2\pi p(\beta + n).$$

The transformation is unaltered if Y is increased to $Y + 2q\pi$, where q is an integer. Thus on $x = a$,

$$Y - y = 2\pi \left(\frac{p}{\alpha + m} + q \right),$$

and on $x = 0$,

$$Y - y = 2\pi [p(\beta + n) + q].$$

If n is chosen such that $(\beta + n)(\alpha + m) \neq 1$, then an infinite number of pairs of values of p and q can be found such that

$$\left(\frac{p}{\alpha + m} + q \right) [p(\beta + n) + q] < 0,$$

and hence either

$$\frac{1}{\alpha + m} > -\frac{q}{p} > \beta + n \qquad \text{or} \qquad \frac{1}{\alpha + m} < -\frac{q}{p} < \beta + n.$$

Since the transformation fulfils the conditions for the theorem, if the theorem is true there will be at least two points which remain invariant under the transformation. Since p and q can take an infinite number of values, if there are two invariant points, there will be an infinite number and hence an infinite number of periodic solutions. Furthermore, the existence of these periodic solutions does not depend on the value of μ.

Poincaré also pointed out that the periodic solution corresponding to a particular pair of values of p and q can only disappear if it coincides with either C_0 or

C_0', in other words if

$$-\frac{q}{p} = \frac{1}{\alpha + m} \qquad \text{or} \qquad -\frac{q}{p} = \beta + n.$$

Birkhoff established the proof of Poincaré's theorem through a *reductio ad absurdum*. His strategy was to show that the assumption that an invariant point did not exist for the transformation led to a contradiction. As Poincaré himself had remarked, if there exists one invariant point, then there necessarily exists a second; hence the theorem fails if it can be proved that there is no invariant point.

If $T(x, y) = (X, Y)$ represents the transformation, then the condition that it has no invariant point is described by

$$(X - x)^2 + (Y - y)^2 > d^2 > 0$$

for all points (x, y) of the ring.

Birkhoff used the coordinate system $x = \theta, y = r^2$, where θ is the angle which a line from the centre to (x, y) makes with a fixed line through the centre, and r is the distance of the point (x, y) from the centre of the ring. The transformation T is then given by

$$X = \phi(x, y), \qquad Y = \phi(x, y),$$

where $X - x$ and $Y - y$ are both single-valued and continuous in the ring R.

He next considered the transformation T_ε defined by

$$X = x, \qquad Y = y - \varepsilon, \qquad (0 < \varepsilon < b^2),$$

which takes the circles $C_a : y = a^2$ and $C_b : y = b^2$ into the circles $C_a' : y = a^2 - \varepsilon$ and $C_b' : y = b^2 - \varepsilon$, respectively. It is then possible to form the auxiliary transformation TT_ε and providing $\varepsilon < d$ this auxiliary transformation TT_ε will also have no invariant point.

If now (x, y) are taken to be the rectangular coordinates of a point in the strip S

$$-\infty < x < +\infty, \qquad b^2 \leq y \leq a^2,$$

corresponding to the ring R, then the transformation TT_ε carries the upper edge of the strip $a^2 \leq y \leq a^2 + \varepsilon$ into the lower edge, and the strip is carried into a second strip lying below the first but with a common boundary. By a repetition of the transformation a series of strips is obtained with eventually the bottom edge of one of the strips, say the kth, overlapping the edge $y = b^2$.

Birkhoff next constructed a curve PQ, where P is a point on C_a, with image P' under TT_ε, and $Q = P^{(k)}$ is the point derived from P by a k-fold repetition of TT_ε, which is the first intersection of the succession of arcs $PP', \ldots, P^{k-1}P^k$ with the lower side of C_b, i.e., Q lies at most ε below C_b (*Figure 7.4.i*). The curve PQ is then invariant under TT_ε.

If a point B moves along PQ, then its image B' under TT_ε will move along the same curve never coinciding with it (since there are no invariant points under the transformation). The angle which the vector BB' makes with the positive direction of the x axis can be taken to be a positive acute angle, and when B has varied to its final position, the same angle lies in the second or third quadrant, since P^k lies to the left of $P^{(k-1)}$ by the hypothesis of the theorem. A rotation of the vector BB' is

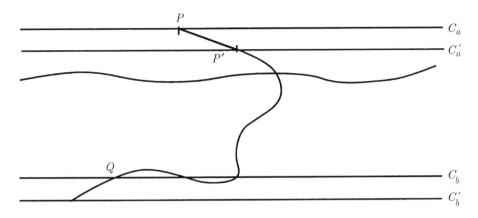

FIGURE 7.4.i

then the least positive angle from the first direction to the second. Furthermore, if
B moves in any manner from a point on C_a to a point on C_b, then the corresponding
vector BB' along the new curve will undergo exactly the same rotation as along
PQ.

Birkhoff considered the inverse transformation $T^{(-1)}$, which is similar to T
except that points on C_a and C_b are moved in the reverse direction. Arguing as
above, if the vector $BB^{(-1)}$ with end point $B^{(-1)} = T^{(-1)}(B)$ has its initial point
B varied from a point of C_a to a point of C_b, then the angle of rotation will be
the least negative angle consistent with its initial and final positions. But the total
rotation of $BB^{(-1)}$ is the same as the rotation of the vector $B^{(-1)}B$ which joins
a point $B^{(-1)}$ to its image under T. Thus by the earlier result, the rotation must
also be the least positive angle, which is a contradiction. Hence there must be at
least one invariant point. To show that there are at least two invariant points it is
sufficient to observe that the total rotation of the vector BB' around the rectangle
$0 \leq x \leq 2\pi$, $b^2 \leq y \leq a^2$ is zero, but around a simple invariant point it is $\pm 2\pi$.
There must be therefore at least two invariant points inside the rectangle.

Birkhoff continued to work on the ideas involved in Poincaré's theorem and in
particular its applications, devoting a chapter to the topic in his general account
of dynamical systems [1927, *150-188*]. In [1925] he extended the theorem to a non-
metric form by removing the condition that the outer boundaries a and $T(a)$ of the
ring R and the transformed ring $T(R)$ must coincide, and replacing it instead with
the condition that a and $T(a)$ are met only once by a radial line $\theta = constant$. In its
revised form he proved that the theorem held for ring-shaped regions with arbitrary
boundary curves and that there are always two distinct invariant points. Since the
extension does not involve an invariant area integral it is essentially a topological
result. Its importance lies in the fact that it can be used to establish the existence
of infinitely many periodic motions near a stable periodic motion in a dynamical
system with two degrees of freedom, from which the existence of quasi-periodic,
i.e., motions which are not periodic but limits of periodic motions, follows.

In [1928] Birkhoff explored the relationship between the dynamical system
and the area-preserving transformation used in the theorem. Having shown that

corresponding to such a dynamical problem there exists an area-preserving transformation T in which the important properties of the system for motions near periodic motions correspond to properties of the transformation T, he then showed that a converse form of this correspondence also exists. In other words, given a particular type of area-preserving transformation, there exists a corresponding dynamical system.

In [1931] Birkhoff generalised the theorem to higher dimensions.

CHAPTER 8

Associated Mathematical Activity

8.1. Introduction

During the period when Poincaré was working on the three body problem and theoretical problems of celestial mechanics, there were other mathematicians and astronomers independently pursuing related topics of research. Mention has already been made of Gyldén and the somewhat unhappy consequences of his priority claim over the discovery of asymptotic solutions, but there were others whose work enjoyed a happier fate.

One mathematician whose work (eventually) met with great success was a Russian working in Kharkov, Alexander Liapunov.[273] While Poincaré had been working on the three body problem, Liapunov had been engaged in a qualitative investigation into the theory of the stability of motion, and in 1892 (some two years after its completion) Liapunov's research was published in a remarkable memoir. Unfortunately, its appearance in Russian meant that its contents were almost inaccessible to the wider mathematical community, and it was some time before his ideas began to penetrate the mathematical circles of Western Europe. For the most part they were initially known only through a series of short notes in the *Jahrbuch über die Fortschritte der Mathematik*, which appeared in 1893, and an extract published in the *Journal de Mathematiques* in 1897.[274] But with the publication of a French translation by Davaux in 1907 (reviewed and corrected by Liapunov), the memoir gradually reached a more extensive audience, and Liapunov's study of the stability question began to be recognised as forming an important complement to that undertaken by Poincaré.

The subject of stability was also taken up in a slightly different way by the Italian mathematician Tullio Levi-Civita. Through his combined interest in geometry, the three body problem and analytical mechanics, Levi-Civita had been led to the study of the qualitative theory of differential equations and associated questions of stability. At the turn of the century he produced a long paper on stability theory [1901] in which he took account of both Poincaré's and Liapunov's ideas and paid particular attention to the restricted three body problem.

With regard to the early responses to Poincaré's researches, these can be broadly divided into two categories. On the one hand, there was the work of those who were engaged in pursuing a solution to the three body problem and for whom Poincaré's memoir was a source of ideas and inspiration; on the other, there was the work of those who had focused on the qualitative aspects of Poincaré's research.

---

[273]For a biography of Liapunov see Smirnov [1992].
[274]*Journal de Mathématiques* (5) **3** (1897), 8.

Of the former, two topics will be discussed here, regularisation and numerical investigations; the latter, which involves the work of Jacques Hadamard and George Birkhoff, will be the subject of the next chapter.

As indicated in Lovett [1912], Poincaré's contribution to knowledge on the three body problem served to generate interest in the problem in several ways, but in the years immediately following the publication of [P2] and [MN I-III], there was one issue which dominated: the regularisation of the equations of motion. It will be recalled that Weierstrass's description of the problem included the assumption that no collisions between the bodies would take place, and Poincaré had based his analysis accordingly. But if a complete solution to the problem was to be found, then collisions had to be taken into account. Since collisions are described by singularities in the differential equations, this raised the question of regularisation. Could the equations be regularised at the points of singularity? This in turn raised a further question about the nature of the singularities. Was it possible that singularities other than collision singularities could exist? If these problems could be resolved then, as Poincaré had indicated [1882a, 1886], it was theoretically possible that a complete solution to the problem could be found. Of the several distinguished mathematicians who applied themselves to these issues, particular consideration is given to the contributions made by Paul Painlevé, Tullio Levi-Civita, Giulio Bisconcini, Karl Sundman and Hugo von Zeipel.

Another aspect of the three body problem which has so far eluded discussion is the question of numerical solutions. Although Poincaré's interest had been in the theoretical side of the problem, to what extent did his results affect the pursuits of those seeking a practical solution? One area in which Poincaré's influence is unquestionably present is the numerical construction of periodic solutions. Work in this field was pioneered by George Darwin, who devoted several years to its study with notable success. Darwin's work is of particular interest, not only because he directly inherited ideas from Poincaré but also because he laid strong foundations for a field of activity which is flourishing more than ever today. The availability of powerful electronic computers has meant that numerical integration is now often a relatively quick and efficient way of gaining an insight into a complex dynamical problem. This is in complete contrast to the painstaking efforts required by numerical analysts at the end of the 19th century, of which Darwin's work was a model of tenacity.

In addition to those mentioned above there were those mathematicians motivated by considerations not necessarily confined to the realms of the three body problem and celestial mechanics but whose study of Poincaré's techniques resulted in further new discoveries in other related fields. Although reasons of space preclude a discussion of their work here, the following is an indication of the kind of research being done.

With regard to the theory of ordinary differential equations, Ivar Bendixson [1901] successfully extended some of Poincaré's ideas concerning the behaviour of solution curves near singularities, and included amongst his results is the theorem, now called the Poincaré-Bendixson theorem, which provides a positive criterion for the existence of a periodic solution in a dynamical system with one degree of freedom.

Notable progress was also recorded in Poincaré's theory of invariant integrals, as for example in the work of Koenigs [1895], who made a connection between it and Lie's theory of contact transformations.[275] At the beginning of the 1920s Elie Cartan [1922] undertook an analysis in which he looked particularly at the relationship between invariant integrals and his own idea of integral forms. In studying the application of the theory relative to the integration of differential equations, Cartan also established the link with Lie's theory of transformation groups.

The selection of topics discussed in this chapter does not claim to form a comprehensive account of mathematics connected with Poincaré's work on the three body problem, and neither is their treatment intended to be an in-depth mathematical analysis. Rather, the objective is to put Poincaré's work into context with some contemporary and later research by both indicating how his work fitted in with corresponding mathematical ideas and giving an insight into the breadth of his influence with regard to the three body problem.

8.2. Stability

As described in Chapter 3, Poincaré first discussed the stability of solutions to differential equations in [1885]. As he had explained, part of his original motivation for developing the qualitative theory had been his desire to tackle the problem of the stability of the solar system, so stability per se was a natural topic for him to pursue. Furthermore, while discussing stability in [1885], he had also been undertaking an investigation into stability of another sort: that of the different forms of rotating masses of fluid. This not only resulted in an important paper [1885b],[276] but it also provided the first connection between his work and that of Liapunov. Only one year before, in 1884, Liapunov had completed his master's dissertation on the subject of the stability of figures of equilibrium, and, as Gray [1992] has pointed out, there were strong similarities between his methods and those of Poincaré; the publication of Poincaré's paper prompted Liapunov to initiate a correspondence between them.[277]

However, in the early 1890s, when each put forward his ideas about the stability of a given state of motion, there were marked differences between the two accounts. Although Liapunov freely acknowledged the influence of Poincaré's [1885] ideas, he used a different definition of stability from that adopted by Poincaré, and while Liapunov developed a precise theory which was quite general in its application, Poincaré's treatment was altogether less rigorous.

Liapunov's memoir, which had originally been prepared as his doctoral thesis, was completed in 1890, but since the process of publication had taken two years, it appeared after Poincaré's accounts in [P2] and [MN I]. It was, therefore, entirely independent of Poincaré's work, although, as Liapunov explained in the preface,

---

[275]See Whittaker [1937, *275*]. For a thorough systematic study of the early theory of invariant integrals, see De Donder [1901].

[276]For an appreciation of Poincaré [1885b] see Darwin [1900].

[277]See the Liapunov-Poincaré correspondence published by V. I. Smirnov and A. P. Youshkevitch in *Cahiers* **8** (1987), *1-18*.

Alexander Liapunov

the delay had given him the opportunity to add notes to the text indicating the analogies with [P2]; [MN I] appeared too late for a similar exercise.

8.2.1. Liapunov's memoir of 1892.
The motivation for Liapunov's research was the desire to ascertain the domain of validity of a certain method of solving differential equations. In this method, in order to obtain a solution, the equations were reduced to linear approximations through the retention of only the first order terms in the dependent variables.[278] This was a considerable simplification of the original equations and especially useful in the case of equations with constant coefficients. But as Liapunov realised, there was no *a priori* reason for such a linearisation procedure to be valid. He therefore set out not only to define the cases where the linear approximations could legitimately be used to replace the original equations but also to derive procedures that would show when the method was invalid. What resulted was a stability theory complete with a battery of techniques designed to deal with a variety of situations.

To make the problem more manageable, Liapunov limited himself to examining only those equations where the coefficients were either constant or periodic, the first case being regarded as a particular case of the second. As he observed, this was not a severe restriction, since, from the point of view of practical applications, many important examples consist of equations of these two types.

Liapunov defined the solution of a system of differential equations as stable if other solutions which started at a given time sufficiently close to the given solution remained arbitrarily close to it at all later times. More formally, he stated that a solution $\Phi(t, \alpha)$ was stable if, given $\varepsilon > 0$, there existed a positive number N (not necessarily an integer) which depended on ε such that

$$|\Phi(t, \alpha + \delta) - \Phi(t, \alpha)| < \varepsilon$$

for all t provided $|\delta| < N$. This was in contrast to the rather freer definition employed by Poincaré (which he had originally used in [1885] and had ascribed to Poisson in [P2]) in which he regarded the motion of a point as stable if it returned infinitely often to positions arbitrarily close to its initial position.

In his determination of stability of a system Liapunov derived two different methods. The first was applicable when it could be presupposed that an explicit solution to the equations of perturbed motion, generally in the form of infinite series, was known. The second (known as Liapunov's *direct* method) could be used when there was no explicit knowledge of the solutions, the method being essentially based on energy considerations due to Lagrange and made rigorous by Dirichlet. Broadly speaking, the method exploits the intuitive idea that an equilibrium state of a physical system is stable if nearby the energy is always decreasing. The stability of the system can then be determined by means of the properties of a certain scalar function positive definite in the domain of the state of equilibrium.[279]

---

[278]Liapunov quotes Thomson and Tait, Routh, and Zukovsky as being the main proponents of this method [1907, *204*].

[279]For a clear and concise account of Liapunov's direct method with applications see La Salle and Lefschetz [1961].

In the case where the coefficients of the equations were constant, Liapunov studied the equations of perturbed motion to discover that the stability of the solution was determined by the roots of a certain eigenvalue equation, these roots being equivalent to what Poincaré was later to call characteristic exponents. He found that if all the roots had negative real parts, stability was guaranteed, and, providing the initial perturbations were small enough, the perturbed solution asymptotically approached the original solution, and the linear approximation could be freely used. If only some of the roots had negative real parts, he found that the system did have a certain conditional stability that could be defined, while if the roots had positive real parts then the system was unstable and the approximation invalid.

Looking at the question from the other way round, his analysis showed that in most cases where the equations had constant coefficients, studying the linear approximation was sufficient to resolve the stability question, the only exception being when some of the roots, without having any positive real parts, had zero real parts. Although these cases were very difficult to analyse, Liapunov recognised, as Poincaré had done, that they were of particular interest, especially if the system of equations was canonical. For in this case, the roots are equal in magnitude but opposite in sign; thus absolute stability is only possible if all the roots have zero real parts.

Since the difficulties in the analysis were determined by the number and type of root generated by the equations, Liapunov made a detailed examination of the two simplest cases: the case in which one root was zero and the case in which two roots were purely imaginary, in each case the remainder of the roots having negative real parts. In both these cases he was able to define the conditions for stability. In the first he found that it essentially depended on the form of a particular series obtained from the equations of the perturbed motion, while in the second he found it was the existence of a periodic solution to the original unperturbed equations which provided the key. Furthermore, as he discovered [1907, *392*], showing how this latter condition worked in practice provided a direct overlap with one of Poincaré's results from [P2]. It turned out that for different reasons and using different methods they had each proved the existence of a periodic solution to a system of nonlinear equations.

In the case of equations with periodic coefficients, Liapunov showed that the stability depended on the roots ρ of another eigenvalue equation related to the roots λ of the previous case by

$$\lambda = \frac{1}{w} \log \rho,$$

where w is the period of the coefficients. In this second case, the stability is determined by the modulus of the roots. If all the roots have modulus less than one then there is stability, and, as in the first case, if the initial perturbations are sufficiently small, the perturbed motion will asymptotically approach the original motion. Similarly, roots with a modulus of greater than one imply instability, and roots with modulus equal to one are the ones which require a more detailed analysis.

Thus while there were certainly many similarities between the sort of results obtained by Liapunov and those obtained by Poincaré, the variance in their definitions of stability meant that the scope of their analysis was substantially different. Liapunov's theory, while extremely rigorous and detailed, was limited in its range by his definition, which is realistically too demanding. For, if a solution is Liapunov stable, not only can the perturbed motion not stray far from the unperturbed motion, but also each point in the trajectory of the perturbed motion has to be close to its contemporaneous point in the unperturbed motion. Since in practical terms there are very few dynamical systems which completely satisfy Liapunov's criteria, the application of his theory is essentially confined to local analysis.

On the other hand, Poincaré's stability theory, being based on a less restrictive definition, could be applied to problems of a far more complex nature than those which could be considered by Liapunov. The point of departure for Poincaré was his theory of invariant integrals, which, in conjunction with his definition, meant that he could attack general questions about the stability of dynamical systems, deriving results such as his recurrence theorem, which allowed him an insight into the behaviour of the solutions of the restricted three body problem. His theory therefore led to knowledge about the global behaviour of systems, knowledge which would have been impossible to obtain within the constraints of Liapunov's theory, although this was to some extent counterbalanced by the accompanying imprecision in his local analysis. Furthermore, Poincaré's ideas about stability provided George Birkhoff with the foundation for his theory of recurrent motion, which is discussed in Chapter 9.

The initial inaccessibility of Liapunov's work meant that it was Poincaré's ideas which met with the first response.[280] However, with the publication of the French translation of Liapunov's memoir, Liapunov's stability theory became more widely known, and the potential of his work began to be recognised. Liapunov's theory, as well as being capable of greater generalisation and having a definition which was intuitively more natural than that of Poincaré, provided a precise and conventional framework within which to work. Today the theory is generally regarded as one of the fundamental achievements within the qualitative theory of differential equations. A substantial literature has grown up around Liapunov's work,[281] most recently in the area of control theory; in 1992 the centenary of the memoir's original publication was commemorated by the appearance of an English translation.[282]

8.2.2. Levi-Civita. An alternative approach to stability theory was put forward by Levi-Civita in a series of abstracts in the *Comptes Rendus* [1900, 1900a, 1900b]. These were brought together in a long paper [1901] in which he placed the new ideas of both Poincaré and Liapunov into the structure of the classical analytical mechanics of Lagrange. As Dell'Aglio and Israel [1989] have eloquently argued, Levi-Civita's work on the qualitative theory of differential equations and related issues of stability provides a convincing example of Thomas Kuhn's "essential tension" between tradition and innovation.

Levi-Civita's interest in classical mechanics was combined with a deep geometrical insight, which meant that the qualitative theory of differential equations formed a natural subject for his research. However, this alignment of mechanics

---

[280]See Gray [1992, *520*].

[281]For an extensive bibliography on the qualitative theory of differential equations and Liapunov stability in particular, see Cesari [1959].

[282]See "Liapunov Centenary Issue", *International Journal of Control* **55** (March 1992), *531-773*.

and geometry led him to a definition of stability which, although very similar to that of Liapunov, differed in one critical respect. While Liapunov's definition only took account of future stability, Levi-Civita's definition incorporated both past and future stability, reflecting the principle of reversibility in physical processes. Another important aspect of Levi-Civita's work is that, in contradistinction to that of Liapunov, it allows for the treatment of the case when the first-order approximation is insufficient. Levi-Civita gave his definition in the following form.

He considered a system of differential equations with periodic coefficients

$$(44) \qquad \frac{dx_i}{dt} = X_i(x_1, \dots, x_n, t) \qquad (i = 1, \dots, n),$$

and said that the periodic solution $x_i = 0$ was stable if and only if for any small neighbourhood E of the origin, there exists a second neighbourhood H such that if the initial position of the moving point is taken in H, the point remains in E for all positive and negative values of t. Although Levi-Civita had opened [1901] by acknowledging Liapunov's definition, when he gave his own definition he made no reference to Liapunov. Later, in his classic work on rational mechanics he referred to his definition as being "absolute stability in the sense of Dirichlet", giving the explanation that it had been derived from a configuration space interpretation of the classic definition of a stable equilibrium.[283]

To deal with the question of the stability of the periodic solutions of equations (44), Levi-Civita employed a geometric model which reduced the question to an analysis of certain point transformations associated with the solutions.

He applied the theory to the restricted three body problem and derived the interesting result that when a periodic solution in the restricted three body problem is such that the mean motions of the planetoid and the other two bodies are commensurable, then the motion is unstable and there will be solutions approaching and receding from the given periodic solution.

With regard to the relationship between Levi-Civita's work and that of Poincaré, although Levi-Civita did not use the notion of Poisson stability, he did explicitly state his agreement with Poincaré's conclusion that instability is the rule and stability the exception.[284] While Poincaré's conclusion specifically related to differential equations of first order and first degree, in Levi-Civita's work it was shown to be true in a more general sense. Furthermore, as Dell'Aglio and Israel have described,[285] there is also a clear parallel between Levi-Civita's geometrical model and Poincaré's method of transverse sections, and from this point of view, Levi-Civita's work can be seen naturally as coming between that of Poincaré and Birkhoff.

8.3. Singularities and regularisation

There were essentially two problems that arose in connection with the singularities of the differential equations of the n body problem. In the first place there

---

[283](With U. Amaldi) *Lezioni di meccanica razionale* **2** (1926/27), *464*. See Dell'Aglio and Israel [1989, *297*].

[284]Levi-Civita [1901, *222*]. See Chapter 3.

[285]Dell'Aglio and Israel [1989, *301-304*].

was the question of determining the types of singularities that could arise and the corresponding investigation of their properties. In this respect, there was not only the subject of collision singularities, which still required a detailed study, having been largely ignored as being of little practical consequence; but there was also the question of the existence of other types of singularities. Once the nature of the singularities had been established, the second problem became the task of eliminating them—the so-called regularisation of the equations.

The three body problem eventually succumbed to resolution on both these issues. The collisions were analysed and a complete knowledge of the type of singularities was obtained. However, resolving these questions for the general n body problem has proved much more elusive, and it is only very recently that significant progress has been made.

8.3.1. The three body problem. In the autumn of 1895 Oscar II once more showed his enthusiasm for mathematics by sponsoring a series of mathematical lectures held at the University of Stockholm. The King had originally intended the lectures to be an extension to his competition and had hoped to entice Poincaré to Stockholm with an invitation to lecture on recent progress in analysis. When Poincaré was unable to accept the offer, the position was filled (on the recommendation of Mittag-Leffler) by another French mathematician, Paul Painlevé.[286] The lectures, inaugurated by the presence of King Oscar himself, were a great success and subsequently gained further acclaim through their publication.[287]

Painlevé, following the brief to lecture on analysis, took as his principal subject the theory of transcendental functions defined by differential equations. In the final part of the lectures he considered the application of his earlier theory to the three and n body problems and investigated the singularities in the differential equations. It was obvious from the equations that a collision point was a singular point, but what was not so clear was whether there was any other type of behaviour which would also lead to a singular point.

In the three body case Painlevé supplied the answer: the only singularities are collisions. More precisely, he stated that starting from an initial time t and given initial conditions, singularities can only occur when at least one of the three mutual distances tends to zero as t converges to a finite time t_1. In other words, either the motion is regular as t increases indefinitely, or there is a collision. What was especially important about the theorem was that it allowed Painlevé to conclude that the equations of motion of the three body problem were integrable using convergent power series (fundamentally equivalent to Taylor series), but only providing the initial conditions were such as to exclude the possibility of a two or three body collision within a finite time.

Thus it was clear to Painlevé that a mathematical solution to the three body problem could be found if it were possible to define precisely the initial conditions which corresponded to a collision. In [1896, 1897] he conjectured that these initial conditions should satisfy two distinct analytic relations (which would reduce to one in the case of planar motion). Then, having made a generalisation of Bruns'

---

[286] *Œuvres de Painlevé* I, *199*. See *Cahiers* **10**, 1989, *194*.

[287] Painlevé [1897]. (Of added appeal is the publication's beautifully lithographed appearance.)

theorem on the existence of algebraic integrals for the three body problem [1897a, 1898], he proved that the relations had to be transcendental [1897b], but further progress eluded him.

As far as singularities of the n body problem were concerned, Painlevé made little headway. He did manage to find a sufficient condition for a singularity to be a collision, but he was still left with the unresolved question of whether *pseudocollisions* (the name he gave to singularities not due to collisions) could exist for $n \geq 4$.

In 1903 Levi-Civita published the first of several papers on regularisation in the three body problem. It was a topic which maintained his interest for over twenty years and which began with two notes in the *Comptes Rendus* [1903, 1903a]. The results in these notes were united in an important paper on the singular trajectories and collisions in the restricted three body problem [1903b]. In this paper he characterised the singular trajectories in the restricted problem, finding the analytic relation predicted by Painlevé.

Levi-Civita formulated the problem by considering the bodies as material points S and J with masses $1 - \mu$ and μ, and planetoid P with negligible mass and putting the equations of motion into canonical form

$$(45) \qquad \frac{dr}{dt} = \frac{\partial F}{\partial R}, \qquad \frac{d\theta}{dt} = \frac{\partial F}{\partial \Theta}, \qquad \frac{dR}{dt} = -\frac{\partial F}{\partial r}, \qquad \frac{d\Theta}{dt} = -\frac{\partial F}{\partial \theta}$$

with the Hamiltonian

$$F = \frac{1}{2}\left[R^2 + r^2\left(\frac{\Theta}{r^2} - 1\right)^2\right] - U + \mu r \cos\theta - \frac{1}{2}r^2$$

where

$$U = \frac{1 - \mu}{r} + \frac{\mu}{\Delta}, \qquad r = SP, \qquad \Delta = JP, \qquad \theta = \angle JSP.$$

The conjugate variables R, Θ are, respectively, the derivative of the radius vector r and twice the (absolute) areal velocity.

As Levi-Civita knew from Painlevé's theorem, every singularity of the motion occurs only when, as t approaches a finite value t_1, the $\lim_{t \to t_1} r = 0$ or the $\lim_{t \to t_1} \Delta = 0$, that is, the differential equations have singularities at S and J respectively. Since S and J enjoy a symmetric role in the problem, it is sufficient to investigate the behaviour of the system about only one of these points, say S, and characterise the singular trajectories Σ along which a collision between S and P can occur.

The motion is regular before $t = t_1$ and so $r \neq 0$ for $t < t_1$. Therefore there exist values of t arbitrarily close to t_1 for which $\dfrac{dr}{dt} = R$ is not identically zero. Using the Jacobian integral and eliminating dt, R can be defined as a function of r, θ and Θ, and the system of equations (45) takes the reduced form

$$(46) \qquad \frac{d\theta}{dr} = -\frac{\partial R}{\partial \Theta}, \qquad \frac{d\Theta}{dr} = -\frac{\partial R}{\partial \theta}.$$

Since along the trajectories Σ there are values of r arbitrarily close to 0 for which θ and Θ are analytic functions of r (since $\dfrac{dr}{dt}$ is not identically zero), these trajectories can be separated from the other solutions to equations (46).

Putting

$$\rho = |\sqrt{r}|, \qquad \theta' = \frac{\Theta}{\rho^4} - 1, \qquad H = -\rho R,$$

where θ' is the relative angular velocity $\dfrac{d\theta}{dt}$, Levi-Civita arrived at the system

(47) $$\frac{d\theta}{d\rho} = -2\rho^2 \frac{\theta'}{H}, \qquad \rho \frac{d\theta'}{d\rho} = -4(\theta' + 1) - 2\mu\rho \frac{W}{H},$$

where $W = \sin\theta \left(1 - \dfrac{1}{\Delta^3}\right)$.

He then proved that the singular trajectories along which S and P collide within a finite time correspond to the single infinity of solutions of (47) which are analytic for $\rho = 0$ and to those alone. If θ_0 is the value of $\theta(0)$ and $\theta'(0) = -1$, then these solutions are of the form

(48) $$\theta = \theta_0 + \rho\alpha(\rho, \theta_0), \qquad \theta' + 1 = \rho\beta(\rho, \theta_0),$$

α and β being power series in ρ.

Hence if a collision takes place, the motion must be along one of these trajectories: that is, it is necessary and sufficient that at each instant ρ, θ and θ' satisfy the equation derived from (48) by eliminating θ_0. Since the first expression in (47) shows that θ_0 is an analytic function of ρ and θ, the second expression can be written

(49) $$\theta' + 1 = \rho f(\rho, \theta),$$

where f is an analytic function of ρ in the domain of $\rho = 0$ for all real θ, and which, as Levi-Civita showed, is a periodic function of θ and can be theoretically determined. He then proved that not only is the relation (49) algebraic in the velocities, periodic and single-valued, but also, as predicted by Painlevé, it is unique. In [1915] he extended this result to the problem of three bodies in a plane.

Encouraged by Mittag-Leffler and Phragmén, Levi-Civita reworked [1903b] in [1906]. He removed the singularities using the transformation defined by

$$x + iy = (\xi + i\eta)^2 \qquad p - iq = \frac{\varpi - i\chi}{2(\xi + i\eta)},$$

which had the advantage of being a simpler transformation than the one he had used in [1903b], as well as being canonical. To regularise the system at the point S, he simply used the auxiliary variable t defined by

$$d\tau = \frac{dt}{\rho^2} \qquad (\rho^2 = \xi^2 + \eta^2).$$

In addition to rationalising his result from [1903b], Levi-Civita was also concerned about its theoretical nature. When the bodies are treated as material points, regularisation only requires the absence of a collision. But from a practical point of view, if the bodies concerned are real celestial bodies, then for the motion to remain regular it is necessary to know not only that there will not be a collision but also that the distances r and Δ will not go below a certain given limit ε. Thus Levi-Civita wanted to establish the initial conditions which would ensure that these mutual distances remained greater than ε.

By regularising the equations he had obtained an analytic representation of all possible arcs A of a trajectory inside a sufficiently small neighbourhood D around S. Since every arc not passing through S remains a finite distance away from S, the minimum distance δ from S to an arc A can be expressed as a function (single-valued inside D) of the initial conditions. Thus either $\delta = 0$ (impact) or $\delta > 0$, and if $\delta > \varepsilon$, then, as Levi-Civita required, physical sense can be given to the mathematical result. However, he realised that was possible for a trajectory to penetrate D infinitely often, leaving along one arc A and re-entering along another arc $A*$, each arc giving rise to a new value of δ (with the possibility of having $\delta = 0$ as a minimum). The problem was then to find out exactly what the lower limit of δ would be. Unfortunately, he was unable to obtain any information about this lower limit, which meant that he could not make long-term predictions about the value of δ. He could only conclude that if in the region D the distance δ is greater than ε, then the motion is regular in the neighbourhood of S; but if the trajectory leaves D and later re-enters again, then it is impossible to forecast its behaviour.

Of special interest in [1906b] is a result which Levi-Civita derived at the end of the paper concerning a solution to the differential equations. When considering the arcs of the trajectories inside the region D, he found a new single-valued solution to the differential equations different from the one given by the Jacobian integral. This was an unexpected result, since it appeared to be in contradiction to Poincaré's theorem on the nonexistence of any new solutions.

However, as Levi-Civita himself explained, there was no contradiction, because the domains of validity of the two results were quite different. Poincaré's theorem established the nonexistence of integrals single-valued with respect to the Keplerian variables. This implied they were single-valued in the neighbourhood of all trajectories which have the same osculating ellipse; whereas Levi-Civita's result implied the existence of single-valued integrals either for only a part of the trajectory or in the neighbourhood of trajectories which are not entirely elliptic.

While Levi-Civita was concerned with defining the conditions for a collision in the restricted three body problem, Bisconcini, a young lecturer at the University of Rome, was working on the same problem but in the general three body case. His results [1906] were published just before those of Levi-Civita [1906b] in the same volume of *Acta*, although they had been completed some two years earlier.

Starting with a system of three bodies P_0, P_1, P_2, with $\rho_1 = P_0 P_1$, $\rho_2 = P_0 P_2$, Bisconcini considered the relative motion of P_1 and P_2 with respect to P_0 and derived the equations of motion in Hamiltonian form. He then concentrated on the case where in the limit as t approaches $t_1, \rho_1 = 0$ and $\rho_2 \neq 0$, i.e., the case of a collision between P_0 and P_1. Making the appropriate change of variable, he arrived at a system of equations, which he called (S), analogous to the equations of motion (47) derived by Levi-Civita in [1903b].

In order to proceed further, he found that he had to make the additional independent assumption that in the neighbourhood of P_0 the angular velocity of ρ_1

in the motion relative to P_0 must be finite. Although he was unable to prove that this was necessarily the case, he had two good reasons for believing it to be true. In the first place, the assumption was known to be true in the restricted problem, and, in the second, as ρ_1 gets progressively smaller, the influence of the point P_2 on the relative motion of P_0 and P_1 tends towards zero, at which point P_0 and P_1 essentially move as a two body system, in which case the angular velocity of P_1 tends towards a finite limit.

Having made this additional assumption, he was able to show that it was possible to put the singular trajectories of the system along which the points P_0 and P_1 collide in a one-to-one correspondence with the solutions of the equations (S) which are analytic in the neighbourhood of the collision. Finally, he deduced two distinct analytic relations between the initial conditions which, when satisfied, proved that the motion was taking place along a singular trajectory, thereby indicating the existence of a collision in finite time.

Bisconcini's result was an important contribution, but it did not provide an altogether satisfactory solution to the problem. In the first place, his solution involved a complicated infinite series (in powers of the distance ρ_1) which was not easy to use. But rather more problematic was the fact that the series was not directly applicable except when the interval of time between the initial instant and the collision was sufficiently short, and he gave no condition for this latter criteria. There was, therefore, still a need both to simplify the solution and to increase the range of its application. Moreover, neither Levi-Civita nor Bisconcini had addressed the question of the conditions for a triple collision.

A complete solution of the three body problem was finally achieved by Karl Sundman, an astronomer at the Helsinki Observatory. Sundman originally published the essential features of his work in *Acta Societatis Scientiarum Fennicae* [1907, 1909] and then later, in response to an invitation from Mittag-Leffler, brought them together in a single memoir published in *Acta* [1912].[288] Not only was Sundman's result quite remarkable, but the methods he used were surprisingly simple. Essentially they depended on the application of Picard's extension to Cauchy's well-known theorem on the existence of solutions to differential equations. The memoir also included a more direct proof of Painlevé's result, as well as a proof of the validity of Bisconcini's postulate concerning the angular velocity of the radius vector in the case of a binary collision.

One of the best known of Sundman's results involves the case of a triple collision. He proved that such an event could occur only if all the constants of angular momentum (areal velocity constants) were simultaneously zero, in confirmation of Weierstrass's conjecture made some twenty years earlier.[289] This then led him to the result that if these constants are not all simultaneously zero and the initial conditions are known, there is a positive limit below which the two greatest of the mutual distances between the bodies cannot go. He further established that if the three bodies collide at the same point in space, they move in the same plane, which

---

[288]Wintner [1947, *428*] observed that Sundman [1909], which included the theory of triple collisions, did not warrant a review in the *Jahrbuch über die Fortschritte der Mathematik* or reproduction in Sundman [1912].

[289]Mittag-Leffler [1912, *58*]. See Chapter 6 above.

Karl Sundman

passes through their common centre of gravity, and as they approach collision they asymptotically approach the equilateral triangle or collinear configuration.

In the case of a binary collision, Sundman showed that the singularity of the differential equations is not essential and so can be removed by a suitable change of variables. Considering the case where the differential equations ceased to be regular for $t = t_1$, he introduced a new independent variable u defined by

$$dt = r\,du \qquad (t = t_0, \quad \text{for } u = 0),$$

from which

$$u = \int_{t_0}^{t} \frac{dt}{t}.$$

where t_0 is a real constant which is chosen in an appropriate way each time the variable u is employed. Thus the system is regular for $u = 0$, and u is known as the regularising variable.

Having introduced u into the equations, Sundman established that the coordinates of the bodies could then be expanded in powers of $(t - t_1)^{\frac{1}{3}}$, and his insight was to realise that an analytic continuation of this expansion could be used to define a continuation of the motion of the bodies after collision. The coordinates then satisfy the differential equations for $t > t_1$ with the same values of the energy constant and the areal velocity constant.

Sundman's description of the motion after a binary collision showed that the orbits of the colliding bodies have a cusp point at the point of collision, whereas the orbit of the third body is continuous in the neighbourhood of the collision. Furthermore, his analysis also showed that the motion can be continued after each new collision, providing not all three bodies collide. In other words, the successive times of binary collision, $t_1, t_2, t_3, \ldots, t_k, \ldots$, cannot have a limit point, so if the sequence of collisions is infinite, then $t* = \lim t_k = +\infty$. Thus the motion can be continued indefinitely for values of t as great as desired.

However, there was a limitation to Sundman's regularisation transformation, $dt = r\,du$, since it was dependent both on the constant t_0 and on whichever of the mutual distances was tending towards zero. To overcome this restriction Sundman introduced another variable, ω, defined by

$$dt = \Gamma\,d\omega,$$

where

$$\Gamma = (1 - e^{-r_0/l})(1 - e^{-r_1/l})(1 - e^{-r_2/l}),$$

so that $\omega = 0$ when $t = 0$, and where the two greater of the mutual distances r_0, r_1, r_2 are greater than l.

Γ has a given value for each real value of the time satisfying $0 \leq \Gamma \leq 1$, and consequently the variables ω and t increase and decrease together. Thus there exists a continuous one-to-one correspondence between the real values of t and the real values of ω, so that when t varies from $-\infty$ to $+\infty$, ω varies likewise.

Given a real and finite value of ω, say ω^*, the coordinates of the three bodies, their mutual distances and the time can be expanded as power series in $(\omega - \omega*)$, where the radius of convergence of these expansions is always greater than a positive limit independent of the value of ω^*, i.e.,

$$|\omega - \omega *| \leq \Omega.$$

The coordinates of the three bodies, their mutual distances and the time are thus analytic functions of ω in a band of breadth 2Ω contained between two lines parallel with the real axis and symmetric with respect to this axis.

Finally by introducing a new variable, τ, defined by

$$\tau = \frac{e^{\pi\omega/2\Omega} - 1}{e^{\pi\omega/2\Omega} + 1},$$

and using the transformation

$$\omega = \frac{2\Omega}{\pi} \log \frac{1 + \tau}{1 - \tau},$$

analogous to the transformation used by Poincaré [1882a, 1886], the band in the ω plane can be transformed into a circle of unit radius in the plane of the new variable τ. The coordinates of the three bodies and the time are now analytic functions of τ everywhere within the unit circle in the τ plane and can be expanded in convergent series in τ for all real values of the time. Furthermore, the same values of l and Ω hold for a group of motions corresponding to different initial conditions, and the different terms of the expansions can be calculated by successive differentiation with respect to τ as soon as the values of l and Ω are determined.

Sundman summarised his achievement in the final theorem of [1912]:

> In the three body problem, if the constants of angular momentum are not all zero and the initial coordinates and velocities of the bodies are given for a finite time, then two constants l and Ω can be found such that by introducing the variable τ instead of the variable t, the coordinates of the three bodies, their mutual distances and the time can be expanded in entire power series in τ, which converge for $|\tau < 1|$ and represent the motion for all time, whatever collisions occur between the bodies, provided that the motion is continued analytically as described above.[290]

Sundman had thus provided a function theoretical proof to the problem which had engaged the minds of many great mathematicians and astronomers since the publication of Newton's *Principia*, a period of well over two hundred years. He had theoretically solved the three body problem. It was a remarkable achievement and all the more so considering the simplicity of his solution: throughout, his analysis depended only on classical results in the theory of differential equations. It is worth recalling that it was only some twenty years earlier that Poincaré had stated that he believed the complete resolution of the problem would require the use of new transcendental functions [P2,6]. Furthermore, Tisserand, in the final volume of his *Mécanique Céleste* published in 1896, had said, "*The rigorous solution of the three body problem is no further advanced today than during the time of Lagrange, and one could say that it is manifestly impossible.*"[291]

Although the significance of Sundman's achievement was certainly recognised by his contemporaries—Mittag-Leffler's encouragement had resulted in the rewriting and publication of his results in *Acta*; in 1913 the French Academy awarded him the *prix Pontécoulant*, doubling the value of the prize; and both Picard [1913] and Marcolongo [1914] wrote enthusiastic reviews—interest in his work was not consistently maintained. In the decade after the publication of the *Acta* paper,

---

[290]Sundman [1912, *178*].
[291]Tisserand [1896, *463*].

minor corrections, simplifications and extensions to his results appeared, the most notable of which was a simplification due to Levi-Civita [1918] which provided a canonical regularisation of the three body problem in the neighbourhood of a binary collision.[292] But from then on and for the next thirty years, Sundman's work seems to have been almost forgotten. Why did such an important and long-awaited result almost fade into obscurity?

First there were the practical limitations of his results. The rate of convergence of the series which he had derived was perceived to be extremely slow, and so for practical purposes the classical divergent series were thought to be more useful.[293] For example, while George Birkhoff enthusiastically embraced Sundman's theoretical achievement—"*It is not too much to say that the recent work of Sundman is one of the most remarkable contributions to the problem of three bodies which has ever been made*"[294]—his verdict on its application was in quite another vein: "*Unfortunately these series are valueless either as a means of obtaining numerical information or as a basis for numerical computation, and thus are not of particular importance.*"[295]

Secondly, the results Sundman obtained furnished no qualitative information about the nature of the motion. He had provided a mathematical solution but not one that revealed general information about the form of the trajectories.

Forty years after the appearance of Sundman's *Acta* paper, interest began to be revived in his work. Jean Chazy [1952] published an appreciation of Sundman's result in which he looked both at the contents of the memoir and at its influence. Although not ignorant of the limitations imposed by the generality of Sundman's result, Chazy's account is a glowing testimonial to its effect on the direction of subsequent research into the three body problem:

> *Already this solution* [Sundman's] *has led to researches into collisions and close approaches between the three bodies, and prompted the study of infinite branches of the trajectories of the three bodies, and the study of motion as time goes towards infinity. Already the determination of singular trajectories has led to substantial results in the representation and the distribution of trajectories of the three body problem—the consideration of which is as necessary as the study of singular points in the study of an analytic function. Without having resolved in one go all the qualitative questions posed by the three body problem, Sundman's solution has given rise to essential progress in the resolution of these questions. And plenty of questions remain open following Sundman's work—just as in the work of Poincaré.*[296]

Chazy himself had extensively researched both the three and n body problems with significant success and so was well qualified to judge Sundman's achievement. In particular he had investigated the long term behaviour of the solutions of the three

---

[292]See also Hadamard [1915] and Birkhoff [1922]. An analysis of the relationship between the work of Sundman and that of Levi-Civita is given in the article by L. Dell'Aglio and G. Israel, "La regolarizzazione delle equazioni del problema dei tre corpi: Levi-Civita e Sundman, due diverse direzioni di ricerca", to appear in *Physis*.

[293]For an explanation of the slow rate of convergence see Saari [1990]. It is of interest to note a challenge to this traditional view put forward by Cesco [1961].

[294]Birkhoff [1927, *260*].

[295]Birkhoff [1920, *53*].

[296]Chazy [1952, *190*].

body problem, making extensive use of both Sundman's regularising variable and Poincaré's theory of invariant integrals. By studying the 12-dimensional phase space defined by the positions and velocities of two of the bodies relative to the third, he provided a classification of the final motions of the problem [1922]. Apart from the bounded and oscillatory motions, he found three different types in which all three mutual distances became infinite: hyperbolic, parabolic, hyperbolic-parabolic; and two different types in which two of the mutual distances became infinite: hyperbolic-elliptic, parabolic-elliptic. In each case the different types were distinguished by the nature of the final velocities of all three bodies.[297]

Chazy's account of Sundman's work was followed in 1955 by a modernised version of Sundman's theorems in Carl Siegel's acclaimed text on celestial mechanics [1971]. Siegel was in no doubt as to the importance of Sundman's results, placing them as one of the most significant developments in the transformation theory of differential equations after the work of Poincaré.

8.3.2. The n body problem. The first person to make an impression on the closing question in Painlevé's Stockholm lectures on whether noncollision singularities exist in the n body problem for $n \geq 4$ was Hugo von Zeipel [1908]. Although von Zeipel ultimately devoted himself to the more practical aspects of astronomy, he had begun his academic career by studying periodic orbits for his doctoral thesis. His interest in singularities almost certainly stems from his stay in Paris (1904-1906), where he studied under both Poincaré and Painlevé.[298] As already noted, von Zeipel was the author of an extensive article on Poincaré's celestial mechanics, which appeared in 1921 in the edition of *Acta* devoted to Poincaré.

Von Zeipel's theorem on singularities in the n body problem (as given by McGehee [1986]) states:

> If some of the particles do not tend to finite limiting positions as t approaches t_1, then one has necessarily
>
> $$lim(t \to t_1)R = \infty,$$
>
> where R is the maximum of the mutual distances.[299]

In other words, a noncollision singularity can occur only if the system of particles becomes unbounded in finite time. At first sight, it would seem that such a singularity is an impossibility, since a particle escaping to infinity in finite time would have to acquire an infinite amount of kinetic energy. However, as Zhihong Xia [1992] pointed out, since the potential energy of the system is not bounded from below, there is no reason why the kinetic energy should be bounded from above. Unfortunately, von Ziepel's work appears to have faded into obscurity. Twelve years after his theorem was published, Chazy [1920] published exactly the same theorem but with no reference to the earlier version, indicating that by that date von Ziepel's work had been forgotten.[300]

---

[297]Further details of Chazy's research can be found in Arnold [1985, *67*].

[298]See McGehee [1986].

[299]McGehee [1986].

[300]McGehee's [1986] paper traces the history of von Zeipel's result and gives a modern version of von Zeipel's proof.

However, despite von Zeipel's result and the interest in it rekindled by Chazy, the proof of whether noncollision singularities exist turned out to be particularly elusive, and definite results have been achieved only recently. Mather and McGehee [1975] made a significant contribution by constructing a solution to the collinear four body problem in which the particles escape to infinity in finite time. This was still not a complete resolution of the problem, since the solution contained infinitely many elastic collisions prior to the appearance of the noncollision singularity.

The question was finally resolved in the affirmative by Xia [1992], who, using "symbolic dynamics",[301] proved the existence of a noncollision singularity for a system of five particles moving in three-dimensional space. Xia's example involves two binary pairs, the particles in the same pair having the same mass, and a particle oscillating between them. The single particle oscillates along a fixed axis, and each binary pair orbits in a different plane at right angles to the fixed axis, each pair rotating in opposite directions. With this symmetric configuration Xia showed that it is possible to set the initial conditions so that the energy gain of the single particle and the corresponding energy loss in the binary pairs is such that all five particles tend to infinity in finite time, and, furthermore, that the example can be modified for $n > 5$.

After Xia's result was announced, and using a different approach, Gerver [1991] proved the existence of a noncollision singularity in a planar $3n$ body problem with n very large. The question for $n = 4$ still remains open.[302]

8.4. Numerical investigations into periodic solutions

8.4.1. Darwin. George Darwin, the second son of Charles Darwin, was elected a fellow of Trinity College, Cambridge, in 1868, and from 1883 held the Plumian Chair of Astronomy.[303] Darwin was very much a traditional applied mathematician whose interest in a problem was stimulated by putting a mathematical hypothesis to the test by way of numerical calculations. Unlike some of his contemporaries, such as Adams and Hill, he was essentially a practical mathematician who, rather than calculating to an exceptional degree of accuracy, took the pragmatic approach, assessing the situation and calculating accordingly.

Darwin was a great admirer of Poincaré's work, and their research had overlapped on an earlier occasion prior to their investigations into periodic orbits.[304] Previously they had both been involved in investigating the figures of equilibrium of a rotating liquid, and a comparison of their work on this topic provides a good illustration of the complementary nature of their approaches to a problem. Although each investigation had resulted in the evolution of a "pear-shaped" figure, Poincaré's analysis had involved a process of evolution forwards, while that of Darwin consisted of working backwards through time.[305] Darwin's study of periodic

---

[301] For further observations about "symbolic dynamics" see the discussion of Morse's work given in the Epilogue.

[302] A discussion of the research into the existence of noncollision singularities of the n body problem is given in Diacu [1993].

[303] For a biography of Darwin see Sir Francis Darwin [1916], and for an assessment of Darwin's scientific work see Brown [1916a].

[304] See Darwin [1900].

[305] Poincaré [1885b], Darwin [1887].

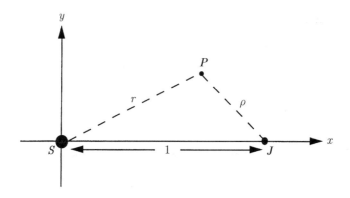

FIGURE 8.4.i

orbits, although owing much to Hill, clearly shows the continuation of his interest
in the work of Poincaré, his orbits giving clear and tangible illustrations of many
of the features previously identified by Poincaré.

Despite the interest in periodic orbits which had been generated by Poincaré's
memoir and the subsequent appearance of the first volume of the *Méthodes Nouvelles*, Darwin's [1897] paper contained the first systematic search for such orbits.
The paper, which had taken him three years to complete, contained the numerical
calculation of periodic solutions of the restricted three body problem, together with
a discussion of their stability. It provided not only extensive details of the numerical results but also a full description of the mathematical methods used to obtain
them.

He derived the equations of motion for the problem using a formulation in which
S, the larger of the two primaries, was placed at the origin of a coordinate system
which was rotating concurrently with the second primary J, with the planetoid P
moving in the plane of J's orbit (*Figure 8.4.i*).

Solving the equations he obtained the Jacobian integral

$$V^2 = 2\Omega - C, \qquad \Omega = \nu \left(r^2 + \frac{2}{r} \right) + \left(\rho^2 + \frac{2}{\rho} \right),$$

where V is the angular velocity of the planetoid, Ω is the overall potential of the
system inclusive of its rotation, and C is the Jacobian constant.

Darwin then took up Hill's idea of partitioning space according to the value of
the Jacobian constant C. Since for real motion $V^2 > 0$, which implies that $2\Omega > C$,
the family of curves $2\Omega = C$ (Hill's curves of zero velocity) define the regions of
space in which the motion of the planetoid is in some way confined. The curves
themselves are the locus of points for which the three bodies move for an instant
as parts of single rigid body.

Darwin considered ρ (the distance between the planetoid and the body J) as
fixed and looked for solutions for r (the distance between the planetoid and the

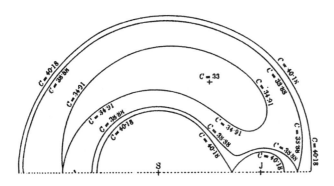

FIGURE 8.4.ii

body S). This led to cubic equations in r, which could be solved to get values of r and ρ to satisfy $2\Omega = C$. Looking at what happened for different values of C, he found that as he changed the value of C the curve went through four critical stages, (α), (β), (γ) and (δ). Each of these stages marked a transition of the shape of the curve, the points in (α), (β) and (γ) being situated on the line SJ, while the points in (δ) were symmetrically placed either side of SJ (*Figure 8.4.ii*). More specifically, at

(α) the internal ovals coalesce to a figure-of-eight, and $r = 1 - \rho$;
(β) the hourglass shape coalesces with the external oval, and $r = 1 + \rho$;
(γ) the horseshoe breaks at the toe, and $\rho = r + 1$;
(δ) C is a minimum, and $r = 1$, $\rho = 1$, and $C = 3\nu + 3$.

In the first three cases Darwin found that the motion at the points was dynamically unstable, while in the last case, which is Lagrange's equilateral solution, he believed the motion to be always stable. (Later the astronomer S. S. Hough [1901] showed that Darwin had made a error with regard to this last conclusion and proved that the (δ) points were only stable for $\nu > 24.9599$, otherwise they were unstable.)

Using the particular value $\nu = 10$ (which is equivalent to a value for the mass parameter μ of $1/11$), Darwin made a classification of the possible periodic orbits depending on the value of C. For example, he found that if C is greater than 40.1821, then the planetoid could either move as a superior planet around both S and J, or as an inferior planet around S, or as a satellite around J; whereas if C is less than 40.182 and greater than 38.8760, then the planetoid could move in the three ways as above, but in addition it could also move in an orbit which incorporated both of the two latter characteristics; or if C is less than 33, there is no region which is forbidden to the planetoid. All these and the other different possibilities can be deduced from *Figure 8.4.ii*.

Darwin included a detailed exposition of his methods of integration, as well as a discussion on the question of stability of the orbits. In the latter he found that he was in agreement with Poincaré concerning the disappearance in pairs of periodic solutions.

As far as actually calculating the orbits was concerned, due to the difficulties involved in discovering periodic orbits making more than one revolution around either of the primaries (or any other point in space), Darwin confined his attention to looking for what he called "simple" periodic orbits. These were periodic orbits which were re-entrant after a single circuit, although they could (and did) include loops. He conjectured that the periodic orbits were the critical cases which separated the orbits into different classes.[306] Thus to find these orbits it was necessary to trace an orbit through its transformation from one class to another.

Due to the possible extent of the field of investigations, Darwin limited himself further by ignoring both the superior planets and retrograde orbits, although later he did make a start on investigating these cases [1909]. In addition, he only considered a range of values for the Jacobian constant between 38 and 40.5.

Darwin's classification of periodic orbits included one family of planets orbiting S; two families of oscillating satellites, each oscillating around a Lagrangian point on the line SJ; and three families of satellites orbiting J. He was, however, struck by the fact that one of these latter families appeared to exhibit a strange characteristic. As the value of the Jacobian constant decreased, the orbit seemed to develop from a simple closed oval into the form of a figure-of-eight in which one loop went round J and the other went round a Lagrangian point on the line SJ. Moreover, accompanying this change of form was a change from stability to instability, a discontinuity which Darwin was unable to explain. Clearly there was an aspect of the behaviour which Darwin's analysis had failed to capture. As mentioned in Chapter 7 above, this seemingly anomalous result attracted the attention of Poincaré, who was able to provide the explanation [MN III, *352*]. There was in fact no anomaly; Darwin had simply been mistaken in classifying the two forms of orbit together when in reality they each belonged to independent families. However, although Poincaré had proved the existence of two different families, his account did not explain the disappearance of the stable orbits and the appearance of the unstable ones. These details were filled in by Hough [1901], who recognised that one of the sources of error was Darwin's failure to take into account that the retrograde orbits were the analytical continuation of the direct orbits. Hough also indicated the existence of another family of figure-of-eight orbits which Darwin himself had not detected.

In the meantime, Darwin had independently discovered his mistake and in [1909] added computational confirmation to Hough's theoretical results.[307] Also included in this second paper are further investigations into the periodic orbits of superior planets, retrograde orbits and orbits of ejection, the latter being those which provide the transitional form between direct and retrograde orbits.

8.4.2. Moulton and Strömgren. Taking their lead from Darwin, two important centres of numerical activity involved in the study of periodic orbits grew up, both making significant contributions to the quantitative analysis of the three body problem.

---

[306] As Szebehely [1967, *434*] has pointed out, Darwin's conjecture, if true, provides a good motivation for studying periodic orbits, but unfortunately its imprecise formulation makes it impossible to establish its validity.

[307] Darwin's appreciation of Hough's contribution is marked by his inclusion of Hough's [1901] paper in his own collected works, sandwiched between [1897] and [1909].

The first of these was in the United States, where, between 1900 and 1917, under the leadership of F. R. Moulton, a research group engaged in both analytical and numerical explorations of periodic orbits prospered. Their work was published in a series of papers assembled in a substantial volume in 1920.[308] With regard to the restricted problem, their numerical work centred largely on locating periodic orbits when the two primaries have equal masses, that is, when the mass parameter $\mu = 0.5$, since this is important in stellar dynamics.

The second, and ultimately the more prolific, of the two research groups was based at the Copenhagen Observatory. Here from 1913 to 1939, Elis Strömgren and his colleagues calculated a comprehensive classification of the periodic orbits for the restricted problem. Again most of their work dealt with primaries of equal mass, with the result that this particular case has now become known as the Copenhagen problem. These calculations also provided the foundations for the celebrated research by Maurice Hénon [1965] into the stability of periodic orbits, which centred on investigating the intersections of the orbits with Poincaré's transverse sections.

Szebehely [1967] gives a clear and concise account of the work of the schools of both Moulton and Strömgren and includes an interesting point about the value of the mass parameter μ. By comparing certain results of Darwin and Strömgren, Szebehely found, contrary to expectation, that the value of μ was not necessarily representative of the magnitude of perturbations, a result which provides an attractive example of the merit of this type of quantitative analysis.[309]

---

[308]Moulton et al. [1920].
[309]Szebehely [1967, *491*].

CHAPTER 9

Hadamard and Birkhoff

9.1. Introduction

In 1896 the prestigious *Prix Bordin de l'Académie des Sciences* was won by Jacques Hadamard while he was professor at the University of Bordeaux. The set topic, which was to improve the theory of geodesics, was motivated by the use of geodesics on surfaces to represent the trajectories of motion in dynamical systems.[310] Hadamard's response resulted in two papers, [1897] and [1898]. The first, which contained most of the material he had submitted for the prize, was primarily a study of geodesics on surfaces of positive curvature, while the second, published after he had moved to the Sorbonne, expanded on ideas proposed in the prize paper and dealt with geodesics on surfaces of negative curvature. Both these papers were characterised by a qualitative analysis inherited from Poincaré. In the first Hadamard appealed to results from classical differential geometry, while in the second, in which Poincaré's influence is strikingly evident, Hadamard's discussion is dominated by topological considerations. Moreover, it was through working on this mathematics directly derived from Poincaré that Hadamard was led to one of his most important and profound ideas: the distinction between "well-posed" and "ill-posed" problems. The latter in effect characterises what we now explore under the label of "chaos".

Hadamard's use of Poincaré's qualitative approach to the theory of differential equations in his Bordin paper provides a powerful illustration of the strength of Poincaré's new methods. Nevertheless, although Hadamard continued to promote Poincaré's ideas in the genre, he himself did no other active work on the topic.[311] This was not the case with George Birkhoff. Poincaré's *Méthodes Nouvelles* provided Birkhoff with inspiration which resulted in an abundance of remarkable research throughout his career. As Oscar Veblen remarked, *"Birkhoff took up the leadership in this field* [dynamics] *at the point where Poincaré laid it down."*[312]

Birkhoff's generalisations and extensions of Poincaré's ideas incorporate a vigorous use of topology, and, as with Poincaré, the periodic motions play a central role in his theory. His ideas were presented with an admirable clarity of exposition, and it was as a result of his efforts that the qualitative theory of dynamical systems emerged as a fully fledged subject independent of its roots in the discipline of

---

[310]For a study of the history of geodesics during the 19th century, see Nabonnand [1995].

[311]In 1920 and 1925, while at the Rice Institute in Texas, Hadamard gave a series of lectures on Poincaré's work, each of which included a discussion of Poincaré's qualitative theory of differential equations. See Hadamard [1922], [1933]. For Hadamard's views on Poincaré's mathematical œuvre, see Hadamard [1912], [1913], [1921].

[312]Veblen [1946, *282*].

Jacques Hadamard

celestial mechanics. The second part of this chapter contains a discussion of three of Birkhoff's early papers on dynamics: [1912], [1915] and [1917].

In the spirit of the previous chapter, the following account is intended only to give an overall view of the different ways that Hadamard and Birkhoff took up and developed some of Poincaré's ideas and thereby provided the foundations for modern dynamical systems theory. As stated earlier, consideration is given predominantly to work produced prior to 1920.

9.2. Hadamard and geodesics

Hadamard's submission for the Bordin prize was the first major paper in which he tackled a subject other than analysis. He had been attracted to the qualitative theory of differential equations through studying Poincaré, and so was well placed to take advantage of the opportunity presented by the Bordin competition to investigate a topic with importance for the qualitative understanding of dynamical problems. In particular he was drawn by Poincaré's idea of the centrality of the periodic motions, and in the context of the theory of geodesics he made an appealing analogy in which he described the closed geodesics as fulfilling the role of a coordinate system to which all other geodesics are then related [1898, 775].

9.2.1. Geodesics on surfaces of positive curvature. Underlying Hadamard's first paper on geodesics on surfaces was the notion of partitioning the surface using properties of the force function of the dynamical system in order to categorise the behaviour of the trajectories of the system.

To illustrate the basic idea Hadamard considered the problem of the motion of a mass particle on a smooth surface of revolution with cylindrical polar coordinates (r, θ, z) under the action of a force function U independent of θ. In general the trajectory of the particle will remain between two parallels of the surface which, through a suitable choice of the constants of integration, can be chosen so that the smaller of the two values of r corresponds to the larger of the two values of U and vice versa. So in the case of a geodesic the two parallels will have the same value of r. In this way a region of the surface is defined in which r and U vary inversely. Thus if the particle is moving on the surface of a sphere, then in general it will pass infinitely often into the lower hemisphere.

More specifically Hadamard considered the motion of a particle moving on a smooth surface under the influence of a single-valued potential function V of the coordinates of the surface. He then partitioned the surface according to the distribution of the successive maxima and minima of the function V.

He began with the system of differential equations

$$\frac{dx_1}{X_1} = \ldots = \frac{dx_n}{X_n} = dt, \qquad X_i = X_i(x_1, \ldots, x_n),$$

in which X_i are analytic functions of the $x_1, \ldots, x_n$, which are regarded as the coordinates of a point M in an n dimensional space E_n. Thus as M describes a trajectory, V will in general have an infinite number of successive maxima and minima. If

$$X(f) = X_1 \frac{\partial f}{\partial x_1} + \cdots + X_n \frac{\partial f}{\partial x_n},$$

then these maxima and minima are described by

(50) $X(V) = 0,$

which represents a manifold of $n-1$ dimensions. If V is a maximum, then $X[X(V)]$ ≤ 0, and if V is a minimum then $X[X(V)] \geq 0$, the former inequality defining that part of the surface (50) where the trajectory passes from the region $X(V) > 0$ to the region $X(V) < 0$, and the latter inequality defining the converse. The boundary of these two parts of the surface is then composed of the points where the trajectory is tangent to the surface (50). Thus, excluding certain exceptional trajectories, each trajectory crosses the surface (50) infinitely often and passes successively into each of the regions determined by the two inequalities. The exceptions occur when the variation in V is always in the same direction. For example, if V is always increasing, then either it becomes infinite or it tends to a limit. Hadamard excluded the first possibility by assuming that V remains finite in the domain in which the x remain finite, but took account of the second with the lemma (now sometimes known as the derivatives theorem[313]):

LEMMA. *If, when t increases indefinitely, the function $V(t)$ tends towards a limit and the first $n+1$ derivatives exist and are finite, then the first n derivatives tend to 0.*

Thus if the function V and its partial derivatives up to third order and the functions X_i and their partial derivatives up to second order are finite as M moves along a trajectory, then either the trajectory crosses the regions of the surface (50) defined by the inequalities infinitely often or it is asymptotic to the boundary of these two regions.

In the particular case of a particle confined to move on a two-dimensional smooth surface, and where the function V has an infinite number of maxima and minima, Hadamard showed that the surface itself could be divided into two regions. The first, which he called the *attractive* region, contains all points of the trajectory where V has a minimum, i.e., it contains an infinite number of distinct parts of the trajectory, each of which is of finite length. The second, which he called the *repellent* region, contains all points of the trajectory where V has a maximum and is a region in which the particle cannot indefinitely remain. The particle passes infinitely often through each of the regions and, consequently, crosses the boundary between the regions infinitely often. If the surface is regular at every point and V is a regular function of the coordinates of the surface, then in the exceptional cases where, after some given moment, the variation in V remains in the same direction, i.e., where there are only a finite number of maxima and minima, the theory shows that either the trajectory remains in the attractive region for an arbitrarily long period or it asymptotically approaches either a point of unstable equilibrium or a closed trajectory.

---

[313]This lemma was also proved independently by both Kneser and Littlewood. See Cartwright [1965, *744*].

The above implies that if the particle describes a geodesic, then the geodesic passes through each of the two regions infinitely often or is asymptotic to a closed geodesic which represents the boundary between the two regions. Thus each geodesic which passes through each of the regions infinitely often cuts the closed geodesic infinitely often. Moreover, when the curvature of the surface is everywhere positive, Hadamard was led to the stronger result that every closed geodesic is cut infinitely often by every other geodesic. In particular, a surface of positive curvature cannot have two closed geodesics which do not intersect—a result clearly demonstrated on the surface of a sphere where every great circle is cut by every other great circle.

A second important result contained both in the Bordin paper and in [1897] was Hadamard's proof of the converse of Dirichlet's theorem on the stability of equilibrium: that a position of equilibrium is unstable if the kinetic energy is not a maximum. In 1895 Kneser had proved the result for the particular case in which the kinetic energy is a minimum, and at the time of the competition a general proof was still believed to be unavailable. However, as Hadamard acknowledged in [1897], he was in fact preceded in the general result by Liapunov [1907].[314]

In the final part of [1897] Hadamard considered the question of the *domain* of a trajectory or a geodesic, and here made explicit use of several of Poincaré's ideas. To explain what he meant by the domain he used the simple example of a geodesic on a surface of revolution. If the geodesic is not closed but oscillates in a strip between two parallels, then following the geodesic in a given direction, the strip is gradually filled out by the geodesic, and hence the strip is the domain of the geodesic.

Hadamard's ideas about the domain stemmed from a direct analogy with the sets introduced by Poincaré in [1885, *142*] and also involved the use of Poincaré's theory of invariant integrals. To define the domain of a given trajectory Hadamard referred to the simplest case of a dynamical system with two degrees of freedom in which a state of the system is represented by the coordinates of a point in a four-dimensional space E_4. He considered the part of a surface Σ in a three-dimensional space E_3 which is crossed infinitely often by the trajectory. If the trajectory is not closed, then the points of intersection are distinct. This set of points admits at least one limit point, and, for increasing values of the time t, the trajectory passes infinitely often through the neighbourhood of this point. By considering all possible surfaces Σ, this gives rise to a closed set of limit points which is then defined to be the domain of the trajectory in the space E_3. Moreover, if the trajectory returns infinitely often to the neighbourhood of an arbitrary limit point, then the set is not only closed but is also perfect. Hadamard used Poincaré's recurrence theorem to show that the trajectories for which this is not the case could be considered exceptional.

Although Hadamard did not obtain many complete results in [1897], he was successful in establishing a new kind of framework from which a cogent theory could be developed. To echo Poincaré's sentiments expressed in his report on the

---

[314] At the time of the competition, Liapunov's paper was only available in Russian and as such unknown to Hadamard. It was only in the following year, when an extract of Liapunov's results was published in the *Journal de Mathématiques*, that Hadamard became aware of Liapunov's work.

competition,[315] the importance of Hadamard's paper lay in the abundance of new ideas it contained and the potential for future research which it provided, a potential which Hadamard himself developed in [1898]. Hadamard had not only boldly followed Poincaré in adopting a strictly qualitative approach to the problem, but he had also demonstrated the power of Poincaré's ideas by showing how several of them could be applied to a particular class of problems.

9.2.2. Geodesics on surfaces of negative curvature.

In terms of actual results Hadamard was much more successful in his second paper [1898], for it contained a full classification of the different types of geodesics that could exist upon surfaces with everywhere negative curvature. As Poincaré [1905a, *39*] remarked, Hadamard, in responding to the Bordin committee's observations concerning future research on geodesics, had provided a complete solution to the problem of geodesics on surfaces of negative curvature. Again Hadamard's approach was qualitative, but this time he focused on the topology of the surface, in particular the order of connectivity which he used to categorise the geodesics.

He began with the hypothesis that the surface of negative curvature consisted of a number n of infinite independent *sheets*, each limited by a curve C and generated by the motion of the curve as it extends to infinity. Each infinite sheet could then be regarded topologically as bounded by the curve C in its initial position and by the same curve as it extended to infinity. Each sheet was therefore topologically equivalent to a circular annulus, or, in other words, each sheet was a doubly-connected surface.

Although the best known examples of such surfaces are the hyperbolic paraboloid and the hyperboloid of one sheet, surfaces of negative curvature can be formed having any number of infinite sheets. There are, as Hadamard observed, surfaces represented by equations of the type

$$z = k \log \frac{\delta_1, \ldots, \delta_m}{\delta_1', \ldots \delta_n'},$$

where k is a constant, and $\delta_1, \ldots, \delta_m; \delta_1', \ldots, \delta_n'$ are the distances projected on the x, y plane of the point (x, y) to the fixed points $P_1, \ldots, P_m; P_1', \ldots, P_n'$ on the plane. This surface has $m + n + 1$ infinite sheets, with m directions on the side of $z > 0$, n directions on the side of $z < 0$, and one direction in the horizontal sense.

As an example, Hadamard included a diagram of the surface

$$z = k \log \frac{\delta}{\delta'},$$

which corresponds to $m = n = 1$ and which has the general form of *Figure 9.2.i*.[316]

Hadamard also made the point that surfaces of negative curvature can have an arbitrary number of holes. As an example, he considered a surface generated by the two hyperboloids $U = 0$, $V = 0$, which cut each other in a hyperbola. The part of surface $UV = \varepsilon$, where $\varepsilon > 0$, in the region $U > 0$, $V > 0$, has negative curvature and has the general form of a surface with two infinite sheets and one

---

[315] *Comptes Rendus* **123** (December 1896), *1109-1111*.
[316] Hadamard [1898, *741*].

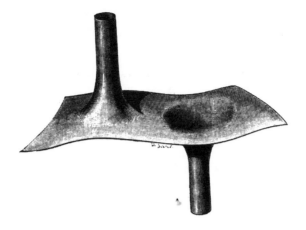

FIGURE 9.2.i

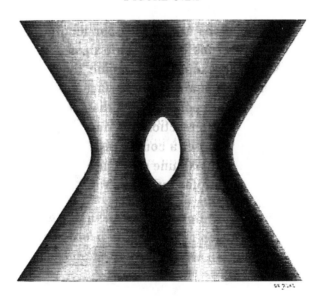

FIGURE 9.2.ii

hole (*Figure 9.2.ii*).[317] It is easy to see that by combining an arbitrary number of similar hyperboloids it is possible to construct a surface with an arbitrary number of holes and two infinite sheets.

Hadamard then investigated the topology of these surfaces by considering the type of curves which could be drawn upon them. He said that two closed curves belonged to the same *species* if they were reducible one to the other by a continuous deformation on the surface. To distinguish between the different species he used two sorts of *elementary* curves: *simple* curves which were equivalent with respect to different edges of the surface and curves which corresponded in pairs to different handles. Furthermore, since any curve could be reduced to a sequence of elementary

---

[317]Hadamard [1898, *744*].

curves in a given direction and order, he had the powerful idea of representing the curves symbolically by using a sequence of symbols, where each symbol in the sequence represented an elementary curve.

To further facilitate his investigation, Hadamard made one more definition. He said that two paths going from a point a on the surface to another point b on the surface were of the same *type* if it was possible to pass from one path to the other by a continuous deformation and the points a and b remained fixed. By this definition, two paths ab, ac, which start from a point a and finish on a curve L, are also of the same type if it is possible to pass from one to the other by a continuous deformation in which the point a remains fixed while the other extremity describes the curve L. If the curve L is closed, then there are an infinite number of ways of going from the point b to another point c of this curve without leaving it, and if the curve can be reduced to a point, then all the arcs are equivalent. In particular, on a doubly-connected surface all the paths from a given point to a given closed curve are reducible one to the other.

In accordance with Poincaré's dictum on the importance of periodic solutions,[318] Hadamard began his study of the different types of geodesics by looking at the closed geodesics. He started by showing that corresponding to each type of line joining two distinct points there is one and only one arc of a geodesic belonging to each type, a result which, as he observed, is equivalent to the theorem that on a surface of negative curvature two infinitely closed geodesics cannot intersect more than once. By proving the impossibility of drawing, either between two points or from a point to a geodesic, two geodesics reducible one to the other, he showed that corresponding to each type of closed curve there is one and only one closed geodesic. He then derived three further results: first, that on a surface of negative curvature there are no reducible closed geodesics; second, that if the surface is doubly-connected, then it only has one closed geodesic; and third, that if the connectivity is greater than two, then the closed geodesics form a denumerable infinity.

Hadamard next considered the distance between two geodesics. It is straightforward to see that there are only three possibilities: either the two geodesics intersect; or the distance between them has a minimum absolute value; or the geodesics approach each other asymptotically. That the first two possibilities exist is evident, but does the third? Clearly, it only makes sense to think of geodesics asymptotically approaching a closed geodesic, but do such asymptotic geodesics exist? To answer this question Hadamard appealed to non-Euclidean geometry. His argument, which proved that such geodesics did indeed exist, went as follows.

Let a' be a point of the surface, and let A be a geodesic joining this point to a point a of a geodesic L. Consider m, which stretches indefinitely along L in a given direction, leaving from the point a and tracing the geodesic $a'm$ of the type A (*Figure 9.2.iii*). The angle $ma'a$ is constantly increasing, but it remains less than a given limit, namely the exterior angle a of the triangle maa'. Thus $a'm$ tends towards a limit L', and the geodesic L' is asymptotic to L. It then follows that corresponding to each type of line joining a point to a geodesic there are two

---

[318]Poincaré [MN I, *82*]. See Chapter 7.

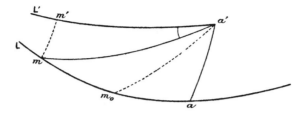

FIGURE 9.2.iii

asymptotes of that particular type. These asymptotes can then be considered as geodesics which join the given point to the points at infinity on the given geodesic.

Since the existence of asymptotic geodesics depend on the existence of closed geodesics, Hadamard classified both types of geodesic together.

Hadamard next focused his attention on funnel-shaped infinite sheets Π_i. Corresponding to each infinite sheet Π_i of this sort is a simple curve and hence a closed geodesic γ_i which can be regarded as the initial position of the curve C_i which bounds the infinite sheet. Furthermore, the theory shows that on Π_i there is only one type of line going from an arbitrary point m to the bounded curve γ_i, hence the geodesic distance u of the point m to this curve is a completely defined single-valued function of the position of the point. Since this distance cannot have a maximum on a geodesic and must increase constantly and indefinitely if it increases at all, a geodesic which enters the sheet Π_i cannot leave it again: it is forced to extend to infinity along the sheet, and, moreover, it must do so regularly (i.e., without returning to a finite distance). As a result, as Hadamard realised, if one geodesic extends to infinity, then so does every geodesic infinitely close to it. He then categorised the geodesics according to whether they were either, first, closed or asymptotic or, second, infinitely extending.

Hadamard next considered the case when all n infinite sheets are funnel-shaped. The corresponding set of n closed geodesics γ_i then divide the surface into n infinite sheets and a bounded part S' which is the *finite part* of the surface. If a geodesic does not extend to infinity regularly on a defined sheet, then, providing the infinite sheet is funnel-shaped, it is always confined in the finite part of the surface. If the sheet is not funnel-shaped, then there are no closed geodesics corresponding to the simple curves, and hence any closed geodesics must be considered as rejected to infinity.

Armed with these results Hadamard drew the following conclusions about the relationship between the geodesics and the order of connectivity of the surface.

If the surface is simply connected then there are no closed geodesics, and the distance from a point M to an arbitrary fixed point O (a distance which is a completely defined function of the position of M) increases indefinitely as M describes a geodesic. Consequently, every geodesic goes to infinity, and the distribution is entirely analogous to that of the lines of a non-Euclidean plane. If the surface is doubly-connected then there is only one elementary curve, the simple curve corresponding to either one of the infinite sheets. The corresponding closed geodesic then divides the surface into two infinite sheets, and there is no finite part of the surface. Hence every geodesic extends to infinity, except for the closed geodesic and its asymptotes, and of the latter there are two through each point of the surface. However, if the order of connectivity is higher than two then not only are the closed curves infinite in number but also the simple curves relative to each infinite sheet are distinct.

Finally, Hadamard found a third category of geodesics quite different from any of those which he had previously described. These geodesics were bounded, but they were neither closed nor asymptotes to closed geodesics. They appeared to approach a closed geodesic asymptotically and then move away before approaching another closed geodesic and so on. Furthermore he found that corresponding to each of these geodesics was an infinite sequence of closed geodesics that got progressively closer together and at the same time progressively increased in length. As he observed, quoting Poincaré:

> ... the geodesic in question possesses the property indicated by Poincaré [MN I, 82], namely that the equations of the problem admit 'a periodic solution (whose period may be very long), such that the difference between the two solutions is as small as desired for any given length of time'.[319]

In other words, Hadamard's discovery of this third type of geodesic provided strong evidence in support of Poincaré's conjecture that the periodic solutions were in fact dense.

Hadamard's proof of the existence of this third and final category of geodesics was especially notable in that it involved an early and novel use of Cantor's theory of transfinite sets.[320] Given a point O and the set E of tangents to the bounded geodesics passing through O, Hadamard discovered that the tangents to the geodesics he had already found, namely the closed geodesics and their asymptotes, were insufficient in number to form the totality of the set E, and hence another category of geodesics had to exist.

Furthermore, he found that in the neighbourhood of a tangent belonging to E there was a geodesic that extended to infinity along an arbitrarily chosen sheet and in the neighbourhood of the same tangent there were geodesics belonging to the third category. Hence the set E was perfect but nowhere dense. This remarkable result brought out an important distinction between the unbounded and bounded geodesics. Earlier he had shown that every unbounded geodesic was surrounded by a continuum of unbounded geodesics, but this result showed that in the case of a bounded geodesic an infinitesimal change in the initial direction of the geodesic was sufficient to make an absolutely arbitrary variation in the final behaviour of the curve. In other words, the boundedness property was not preserved by such a change.

This phenomena of sensitivity to initial conditions led Hadamard to propose the idea of the "well posed problem". Since in reality it was never possible to measure

---

[319]Hadamard [1898, 768-769]

[320]In his discussion of the formation of the set E, Hadamard not only made reference to the work of Cantor and the similar sets encountered by Poincaré, but he also acknowledged an earlier contribution by Bendixson, although without providing a reference [1898, 771].

the initial data completely, he reasoned that it did not make sense to ascribe physical validity to a solution unless the solution had continuity with respect to the initial data. In a discussion of [1898] written only three years later he concluded:

> *Above all it must be acknowledged that the behaviour of these trajectories* [geodesics] *may depend on arithmetical discontinuous properties of the constants of integration. Secondly, as a result the important problems of celestial mechanics, such as the stability of the solar system, may belong to the category of ill-posed problems. If we substitute the search for the stability of the solar system with the analogous question related to geodesics of surfaces with negative curvature, we establish that each stable trajectory can be transformed, by an infinitely small variation in the initial conditions, into a completely unstable trajectory extending to infinity, or, more generally, into a trajectory of any of the types given in the general discussion: for example, into a trajectory asymptotic to a closed geodesic. But, in astronomical problems the initial conditions are only known physically, that is to say with an error which can only be reduced by improving the means of observation but which cannot be eliminated. However small it is, this error might cause a total and absolute perturbation in the result.*[321]

9.3. Birkhoff and dynamical systems

Birkhoff's deep study of Poincaré's work on dynamics is evident from his first publication devoted to theoretical dynamics [1912], in which he introduced the idea of "recurrent motion" as a natural extension of periodic motion. This was followed by his resolution of Poincaré's last geometric theorem [1913] (described in Chapter 7) and two prize-winning papers, [1915] and [1917], the first on the restricted three body problem and the second on dynamical systems with two degrees of freedom. Many of the essential ideas from these early papers are collected together in his acclaimed book *Dynamical Systems* [1927], which was based upon his Colloquium Lectures delivered before the American Mathematical Society in 1920 and in which the influence of Poincaré's work on celestial mechanics is abundantly clear.

9.3.1. Recurrent motion. Birkhoff's first paper on dynamics [1912] marked the beginning of a new phase in dynamical theory. Not only did Birkhoff explicitly consider a general dynamical system as opposed to addressing a particular dynamical problem, but he also thought in terms of "sets" of motions—more specifically "minimal" or "recurrent" sets of motions—as opposed to thinking solely in terms of a particular type of motion.

He began with the general class of dynamical systems defined by the differential equations

$$\frac{dx_i}{X_i} = dt \qquad (i = 1, \ldots, n),$$

---

[321] Hadamard [1901, *14*]. As Mawhin [1994] has observed, the fundamental importance of the physical implications of Hadamard's result concerning the sensitivity to initial conditions was recognised by Duhem [1962, *138-141*] in 1906.

George Birkhoff

where the X_i are n real analytic functions and a state of motion can be represented by a point P in a closed n-dimensional manifold. A motion can then be represented by a trajectory in the manifold, and its domain is its closed set of limit points. He called the limit points of a motion α- and ω-limit points as t becomes negatively or positively infinite respectively and defined a stable motion as one which never became arbitrarily close to a singularity of the manifold.

If M' is a closed set of limit motions (i.e., trajectories composed of limit points) of a motion M and M' contains no proper subset, then Birkhoff called the members of the set M' recurrent motions and the set itself he called minimal. More specifically he proved that a motion is recurrent if and only if for every $\varepsilon > 0$ there exists an interval of time T so large that the arc of the trajectory corresponding to the motion during the interval has points within a distance ε of every point of the trajectory. In other words, a motion is recurrent if during a sufficiently long time interval T it comes arbitrarily close to all its states of motion. A recurrent motion is therefore stable in the sense given above.

A direct connection can be made between Birkhoff's concept of recurrent motion and Poincaré's ideas on stability. By definition every point on the trajectory of a recurrent motion is a limit point, hence the motion must approach every point on the trajectory infinitely often and arbitrarily closely. In other words, every recurrent motion must be Poisson stable. Clearly the simplest type of recurrent motions are the stationary and periodic motions, since in each case M' coincides with M and contains a single motion, whereas in every other case M' contains an infinite number of motions.

With regard to the general problem of determining all possible motions in a dynamical system, Birkhoff highlighted the value of the idea of recurrent motion by two particular results. On the one hand he proved that the set of limit motions of any motion contains at least one recurrent motion, while on the other he showed that any point P either generates a recurrent motion or generates a motion which approaches with uniform frequency arbitrarily close to a set of recurrent motions. The concept of recurrent motion can be used to derive definite results about the motion in an arbitrary dynamical system, and one of the significant features of the theory is that it is valid for systems with any degree of freedom. This is in contrast to Poincaré's theory of periodic motion, which is only known to be valid for systems with two degrees of freedom.

Birkhoff also made a connection between recurrent motion and Hadamard's classification of geodesics on surfaces of negative curvature. It will be recalled that Hadamard's final category of geodesics contained those bounded geodesics which asymptotically approach a closed geodesic and then move away before returning to approach another closed geodesic. Since the motion corresponding to a geodesic of this type is necessarily stable, there must exist at least one geodesic corresponding to a recurrent motion. This cannot be either an asymptotic geodesic or an unbounded geodesic, and hence either every geodesic in the third category approaches infinitely often and arbitrarily closely one particular closed geodesic or there exists a recurrent motion given by a geodesic from Hadamard's final category.

In the case where the differential equations ceased to be analytic Birkhoff proved that recurrent motions do exist but they are not periodic, nor do they occur in the immediate neighbourhood of any periodic motion. Hence he called them *discontinuous* recurrent motions. The question of whether discontinuous recurrent motions existed in the analytic case was resolved by his student Marston Morse.[322]

9.3.2. The restricted three body problem.

Birkhoff's first paper, which concerned the restricted three body problem, was his acclaimed proof of Poincaré's last theorem [1913], which appeared shortly after his paper on recurrent motion. With regard to his treatment of the problem in its generality, he published three principal papers, [1915], [1935] and [1936], only the first of which is discussed here. This paper, for which he won the Quirini Stampalia prize of the Royal Venice Institute of Science, provided the first major qualitative attack on the problem since Poincaré. But, unlike Poincaré, Birkhoff made little concession to analysis, and his investigation was founded almost entirely on topological ideas.

Birkhoff formulated the problem in the standard way using a rotating coordinate system (x, y) with the two primaries S and J and located on the x axis with the origin at their centre of gravity (see *Figure 2.2.ii*),[323] and, as customary, reduced the system from fourth to third order by considering only the totality of motions for which the Jacobian constant C had a given value.

He first established a transformation of the variables which not only allowed him to derive Levi-Civita's [1906] equations and hence remove one of the singularities, but also enabled him to derive a new form of the equations in which the singularities at both S and J were simultaneously removed and in which the equations were regular providing the planetoid was not rejected to infinity. From this new form of the equations he created a geometric representation in which the manifolds of motion were represented by the stream lines of a three-dimensional flow and were without singularity unless C took one of five exceptional values. When these five values were excluded, the totality of the states of motion could then be represented by the stream lines of a flow occupying a nonsingular manifold in a four-dimensional space. Moreover, since the five singular values mark the positions where two of the manifolds are about to join or separate, they act to distinguish between six different manifolds according to the value of C. By considering the representation from a topological point of view, Birkhoff demonstrated that each of the six manifolds required a different model, thereby effectively illustrating the problem's dependence on the value of C.

Birkhoff restricted his research mostly to the case in which the planetoid was confined to move inside an oval about one of the primaries. This case, which is the simplest of the six, occurs when C is sufficiently large and positive. The relative simplicity of the case is associated with the fact that, providing C is sufficiently large (or μ is sufficiently small), the restricted problem closely resembles a two body problem. In this case Birkhoff's topological model shows that the states of motion are in one-to-one continuous correspondence with the points of a sphere, providing the diametrically opposite points are taken as identical.

---

[322]Morse [1921a]. See the Epilogue.

[323]As Szebehely has observed [1967, *37*], the formulation of the restricted three body problem seems peculiarly prone to error. In Birkhoff's formulation his description of the problem, his diagram, and one of his equations all contradict each other.

If the motion is unperturbed the planetoid moves in a rotating ellipse with semi-major axis a, where the minimum value a_1 of a corresponds to a retrograde circular orbit and the maximum value a_2 corresponds to a direct circular orbit. To consider the structure of the manifold of motion Birkhoff chose the variables to be the semi-major axis a, the longitude θ of the line of apsides (the line joining the perihelion and the aphelion) of the ellipse with respect to the rotating x axis, and the mean anomaly ϕ of the planetoid taken in the same direction as the motion. The variable θ then determines the instantaneous position of the ellipse while the variable ϕ determines the position of the planetoid on the ellipse. Thus, providing the circular orbits are excluded (since in this case θ and ϕ are undetermined), the totality of states of motion can be represented in a one-to-one continuous way upon the interior of the hollow cylinder $a_1 < a < a_2$, $0 \leq \phi \leq 2\pi$, and the trajectories are spirals on the cylinders $a = constant$. The direct circular orbits are represented by the outer cylindrical surface $\theta + \phi = constant$, and the retrograde circular orbits are represented by the inner cylindrical surface $\theta - \phi = constant$.

However, as Poincaré had shown [MN III, *372-381*], the representation of the problem as a three-dimensional flow can be reduced to a representation which depends on the transformation of a two-dimensional ring into itself. For if K is any point on the ring $\phi = 0$ in a given cylinder and L is the point at which the trajectory through K first meets the ring $\phi = 2\pi$, then corresponding points on the two rings will represent the same state of motion of the planetoid. In this way the two ends of the hollow cylinder are identified to form a torus with an internal hole of radius a_1. Thus the transformation T which takes K into L defines a transformation of the ring $\phi = 0$ into itself. Moreover, T preserves certain essential properties of the trajectories. For example, if the trajectory is periodic then a certain number of applications of T will take the point K into itself.

Birkhoff therefore constructed in the x, y plane a series of concentric retrograde (direct) circles of radius less than a_1 (a_2) about the primary. These can then be regarded as generating a ring of two leaves joined at the origin in the x, y plane. For any point K of a circle of the ring, there exists a positively tangent orbit which will again become positively tangent at a point L for the first time. This establishes a one-to-one continuous transformation of the ring into itself taking any point K into its corresponding point L, leaving radial distances unchanged, and regresses each point by an amount dependent on the ellipse of motion of the planetoid. He then proved that for sufficiently small values of μ and any value of $C > \sqrt[3]{32}$, it is possible to analytically continue the direct and retrograde periodic orbits and hence generalise the ring construction for the perturbed motion. In this case the transformation is a precise generalisation of the transformation for $\mu = 0$. It is one-to-one, continuous along the boundaries and varies continuously with μ from the transformation for $\mu = 0$. He further proved that a necessary and sufficient condition for the existence of a ring bounded by a retrograde periodic orbit and a direct periodic orbit and which is cut by all the stream lines infinitely often is that every orbit in a sufficiently large time interval makes an arbitrary number of positive circuits of the retrograde orbit.

Birkhoff also showed that the transformation T possesses two important properties, each of which can be used to form a basis for further results. First, T is an area-preserving transformation. Birkhoff used this property to show that for a given value of C and a given point in the plane there are an infinite number of stream lines of the flow which pass through the point at a later time. Second, T is a product of two involutory transformations, and from this property Birkhoff proved the existence of an infinite number of symmetric periodic orbits and also deduced results concerning their characteristic properties and distribution.

One other problem which Birkhoff addressed in [1915] concerned the restriction on the value of μ which was inherent in his method for determining both the retrograde and direct periodic orbits. In the case of retrograde orbits he found that he could remove the restriction by returning to the differential equations of the problem and employing a transformation similar to the one used by Levi-Civita [1906]. In this way he was able to establish that for an arbitrary value of μ there exists at least one retrograde periodic orbit symmetric with respect to the x axis which makes a single orbit around the primary, providing the value of C is such that the planetoid is confined to move within a closed oval of zero velocity about the larger of the two primaries. The case of direct periodic orbits was rather more difficult and required a different approach. In this case Birkhoff sought a new transformation whose construction was based on the existence of a retrograde orbit already known to exist, rather than, as above, a transformation whose construction was based on the existence of both retrograde and direct orbits. The transformation he used was one in which the variables (a, θ, ϕ) are replaced by

$$a* = \frac{1}{a} - \frac{1}{a_2}, \qquad \theta* = \theta, \qquad \phi* = \phi - \theta,$$

and Poincaré's ring transformation is replaced by the transformation T^* of a disc into itself whose only boundary corresponds to the retrograde periodic orbit. A correspondence is then set up between the retrograde orbit for $\mu = 0$ and the retrograde periodic orbit for $\mu \neq 0$. By Brouwer's fixed point theorem, the transformation T^* necessarily possesses an invariant point, and this point corresponds to a direct periodic orbit making a single revolution about the primary. Thus, providing the conditions for the existence of T are fulfilled, there exists a transformation T^* and at least one direct periodic orbit.

Twenty years elapsed before Birkhoff's next publication on the problem. In the interim he had researched prodigiously into general dynamical systems, the crowning result of which was another prize memoir [1935a]. In his two later papers on the restricted problem, [1935] and its continuation [1936], Birkhoff combined his ideas from [1915] together with some of the general results from [1935a], notably his development of Poincaré's idea of a transverse section. In [1935] he focused on the analytic properties of the transverse section and the transformation T used in [1915], while in [1936] he used qualitative methods to explore the results from [1935] in order to obtain further information about the different types of motion and the relationships existing between them.

Birkhoff also contributed to another paper that concerned the restricted three body problem. In 1922 the National Research Council Committee on Celestial Mechanics, of which he was a member, drew up a report on the state of celestial mechanics [1922]. Amongst other topics the report contained a short discussion on the restricted three body problem, and there seems little doubt that Birkhoff was the author of this part of the report. In it were set out the particular difficulties associated with the problem, and it contained an admirably concise and accessible explanation of Poincaré's method of reducing the problem to the transformation of a transverse section.

9.3.3. Dynamical systems with two degrees of freedom.

One of Birkhoff's most important papers on dynamics, and one which was clearly inspired by Poincaré, was his famous paper on dynamical systems with two degrees of freedom [1917], which won the Bôcher Prize of the American Mathematical Society in 1923. According to Morse, Birkhoff declared a few years later that he thought [1917] as good a piece of research as he would be likely to do.[324]

As previously indicated, the interest in dynamical systems of this type stems from the fact that they represent the simplest type of nonintegrable dynamical problems. Thus, as exemplified in the work of Poincaré and Hadamard, they form the natural starting point for qualitative explorations into questions of dynamics. Furthermore, as Birkhoff showed in [1917], they have the advantage that it is always permissible to consider the motion as the orbit of a particle constrained to move on a smooth surface.

Birkhoff began with the equations of motion in standard Lagrangian form

$$\frac{d}{dt}\left(\frac{\partial L}{\partial x'} - \frac{\partial L}{\partial x}\right) = 0, \qquad \frac{d}{dt}\left(\frac{\partial L}{\partial y'}\right) - \frac{\partial L}{\partial y} = 0,$$

where the function L, which is quadratic in the velocities, involves six arbitrary functions of x and y. By making an appropriate transformation of variables he reduced the equations to a normal form, which involved only two arbitrary functions of x and y.

In the *reversible* case, i.e., when the linear terms in the velocities are lacking in L so that the equations remain unchanged when t is replaced by $-t$, the transformation was already well known. In this case the equations of motion can be interpreted as those of a particle constrained to move on a smooth surface and the orbits of the particle interpreted as geodesics on the surface. But in the *irreversible* case, as for example in the restricted three body problem, Birkhoff's transformation was new, and he gave a dynamical interpretation in which the motions can be regarded as the orbits of a particle constrained to move on a smooth surface which rotates about a fixed axis with uniform angular velocity and carries with it a conservative force field.

The central part of the paper concerned various methods by which the existence of periodic orbits could be established. In the first place, Birkhoff considered a method which he called the *minimum method*. Briefly, given a certain type of Lagrangian dynamical system and any closed curve l not deformable to a point on the surface, then for a given value of the energy constant there exists a periodic orbit of the same type as l for which a certain integral is a minimum. In the reversible cases where the surface is closed and of positive genus, then this integral corresponds to the arc length on the surface and the periodic orbit corresponds to a closed geodesic. In other cases, Birkhoff showed that the knowledge of boundaries of a particular type was required before the method could be used. In the irreversible case the situation is different because the integrand of the integral is no longer of

---

[324]Morse [1946, *380*].

one sign. As Birkhoff observed, the method only yields the completely unstable periodic orbits and hence has a limited application.

Another method developed by Birkhoff was his new *minimax* method.[325] This is applicable only to the reversible case and establishes the existence of a large and completely different class of periodic orbits. As Birkhoff described in [1927, *133*], this method can be most easily understood in an informal way by considering a torus in three-dimensional space. Clearly the minimum method will yield a closed geodesic having the type of a closed curve not deformable to a point on the surface of the torus. If now a closed curve of the same type is moved away from the minimising geodesic while at the same time one of the angular variables defining the torus is increased by $2n\pi$, then the length of the closed curve will increase during this motion, and the curve will have to reach a least upper bound in length in order for the motion to be possible. When the curve reaches this least upper bound length it will be taut, and this position of the curve is the one which corresponds to a closed geodesic of minimax type.

Thirdly, Birkhoff considered Poincaré's method of analytic continuation, which is applicable to both reversible and irreversible periodic orbits. One of the problems with the method was that it was only valid for a small variation in the value of the parameter. The restriction was due to the possibility that the period of the orbit under consideration might become infinite. Thus to increase the interval of the variation it is necessary to show that this possibility cannot arise, and Birkhoff did precisely that for a wide range of periodic orbits.

Finally, it was in this paper that Birkhoff first began to generalise Poincaré's idea of a transverse section and formally develop a theory attached to it. Poincaré had used the idea specifically to reduce the restricted three body problem to the transformation of a ring to itself, but if the method was to have a general validity it was important to establish under what circumstances transverse sections exist. Birkhoff was able to show that not only do they exist in a wide variety of cases but also that they can be of varying genus and have different numbers of boundaries.

Concerning the transformation of a transverse section, Birkhoff emphasised the fact that it possesses an invariant area integral. This is important because it means that the transformation only involves one arbitrary function of two variables as opposed to the normal form of the differential equations, which involves two arbitrary functions. In other words, reducing a dynamical problem to a transformation of a transverse section into itself is both a qualitative and an analytic reduction.

By considering the invariant points of these transformations Birkhoff derived two important results about the periodic orbits. In the first place he found that the difference between the number of unstable and stable periodic orbits is a constant, the constant depending only on the nature of the transformation and the genus and number of boundaries of the original surface. The second result involved a modification to Poincaré's last theorem. As described, the theorem involves the use of a ring-shaped transverse section to prove the existence of an infinite number of periodic orbits. Birkhoff now proved not only that the theorem can be used

---

[325]As noted by Birkhoff [1927, *139*] and Veblen [1946, *283*], Birkhoff's minimax method provided a starting point for Morse's work on calculus of variations in the large, which introduces topological considerations into analysis.

to establish the same conclusion for a general transverse section of genus $p = 0$, but also that a modified version of the theorem can be used to establish the same conclusion for a transverse section of genus $p > 0$.

9.3.4. Later papers.
Birkhoff's work on dynamical systems continued throughout his life, and mention has already been made of his later papers which relate to Poincaré's last theorem and his research on the restricted three body problem. The following brief summary of some of his other work gives an indication of the extent and direction of his later research.

In 1920 Birkhoff published a major paper on *Surface transformations and their dynamical applications* [1920], which was essentially an extensive elaboration and extension of some of the ideas he had broached at the end of [1917]. He began with a classification of the different types of invariant points and considered the behaviour of points in the neighbourhood of invariant points under the one-to-one, direct, analytic transformation of an analytic surface into itself. He also considered the problem of determining the behaviour of other different classes of points. The dynamical applications, which involved questions of integrability, stability and the classification of different types of motion, were only made briefly at the end of the paper.

In [1927a] Birkhoff considered the question of stability and the role of the Hamiltonian form of the equations. He argued that essentially the *only* significance of the Hamiltonian equations is that they possess the property of complete formal stability, i.e., any set of n equations possessing complete formal stability at a point of equilibrium can be given Hamiltonian form by an appropriate change of variables.

In [1927b] Birkhoff considered the distribution of periodic motions in dynamical systems with two degrees of freedom. As an example of how Poincaré's theorem and his own generalisation [1925] could be applied, he discussed the motion of a billiard ball on a convex table. He showed that if such a dynamical system admits a stable periodic motion, then it admits an infinite number of other stable periodic motions within its immediate vicinity, and the totality of these stable periodic motions forms a dense set. Although he did not resolve Poincaré's question of whether the periodic motions are densely distributed throughout the possible motions, his example did show that this cannot be true unconditionally.

Finally, it is appropriate to mention Birkhoff's famous "ergodic theorem" [1931a], since its origins relate back to Poincaré's recurrence theorem.[326] The proof of the theorem is one of the major mathematical achievements of the period and ranks in Birkhoff's œuvre alongside his proof of Poincaré's last theorem.

The ergodic theorem states that for any dynamical system given by differential equations which possesses a certain n-dimensional invariant integral, there is a definite time probability p that any moving point, except those of measure zero, will be in an assigned region. Birkhoff's proof combined Poincaré's topological

---

[326]The term "ergodic" originated with Boltzmann, who used it to describe mechanical systems which had the property that each particular motion when continued indefinitely passed through every configuration and state of motion of the system that was compatible with the value of the total energy.

approach with the use of Lebesgue measure theory. Birkhoff himself gave a clear explanation of the theorem and the nature of its applications in [1942].

CHAPTER 10

Epilogue

10.1. Introduction

The discussions in Chapters 8 and 9 of the impact made by Poincaré's memoir during the three decades following its publication have revealed a somewhat two-sided picture. On the one hand Poincaré's memoir stimulated interest in different aspects of the three body problem, as exemplified in the work of Painlevé, Levi-Civita, Sundman and Darwin, but, on the other, Poincaré's innovative topological approach to dynamics, although lauded by his contemporaries, found little expression in their work. Hadamard, despite the success of his Bordin paper, did not use it as a focus for later research, and Birkhoff, although he made remarkable progress with the development of Poincaré's ideas, belonged to a younger generation of mathematicians, and his investigations did not begin until 20 years after the publication of Poincaré's memoir. Furthermore, certain issues raised by Poincaré's work, such as the convergence of the series used in celestial mechanics and the strange characteristics exhibited by his doubly asymptotic solutions, were not the subject of any extensive research during the period under review. Much more recently each of these topics has resurfaced in the foundations of important new branches of mathematics, the former in Kolmogorov-Arnold-Moser (KAM) theory and the latter in the theory of chaos. This prompts questions about why these issues were not raised earlier and why there was a general lack of contemporary response to Poincaré's dynamical ideas.

It is conspicuous that during the early years of the 20th century no serious attempt was made to investigate further the behaviour of Poincaré's doubly asymptotic solutions, the complexity of which Poincaré so forcefully described in the final volume of the *Méthodes Nouvelles*, published in 1899. To a large extent this can be explained by the inability of researchers to engage in a quantitative analysis due to inadequate computing techniques. The advent of the modern digital computer has meant that such an analysis is now possible, with the result that the last twenty years has seen an explosion of research into nonlinear systems, producing a wealth of new results. One well-known consequence of this has been the unfolding of the mathematical theory of chaos, the very heart of which lies in Poincaré's theory of doubly asymptotic solutions.[327]

In addition to the problems of numerical computation, there was also the difficulty caused by the fact that the apparently random behaviour exhibited by Poincaré's doubly asymptotic solutions did not fit in with the then widely accepted

---

[327] For a mathematical description of Poincaré's work and the origins of chaos theory see Goroff [1993]; for informative popular accounts see Ekeland [1988], Stewart [1989] and Holmes and Diacu [1996].

Laplacian model of a clockwork universe. Indeed, as observed earlier, it may be that this belief in some kind of ultimate order was partly responsible for Poincaré himself missing the chaotic behaviour in his original memoir.

The idea of inherent chaos has proved a difficult concept to grasp, and much of the progress that has been made with this element of Poincaré's work is relatively recent and beyond the scope of this book. The first milestone in the theory's development was the work done by the Russian mathematician V. K. Melnikov. Melnikov, who had closely studied Poincaré, explored questions relating to Poincaré's asymptotic solutions and developed an analytical method for the detection and investigation of chaotic motions within a dynamical system.[328] Melnikov's method, now an extensively applied tool, is based on Poincaré's criteria for the continuation of periodic and homoclinic solutions and is used for studying the first return map for specific examples of periodically disturbed differential equations. Where Poincaré used power series approximations to the Hamilton-Jacobi equation to prove that the stable and unstable manifolds split and intersect transversely, Melnikov calculated the perturbed solutions using the globally computable solutions of the unperturbed integrable system.[329]

Briefly, in the simplest case, Melnikov's method can be applied to a periodically forced system of the form

$$\frac{dx}{dt} = \phi_1(x, y) + \varepsilon\Psi_1(x, y, t),$$

$$\frac{dy}{dt} = \phi_2(x, y) + \varepsilon\Psi_2(x, y, t).$$

When it is unperturbed ($\varepsilon = 0$) the system is integrable, since it is then a one degree of freedom Hamiltonian system. Assume that the unperturbed system possesses a homoclinic solution connected to a fixed saddle point p_0, and since the perturbed system is time dependent, rewrite it as an autonomous system in three dimensions. The point p_0 then corresponds to a periodic solution with two-dimensional stable and unstable manifolds which intersect along a two-dimensional homoclinic manifold Γ_0. Now define the two-dimensional transverse section of the phase space and construct the Poincaré map of the transverse section into itself. Consideration of the unperturbed and perturbed Poincaré maps of the system shows that the fixed point p_0 is perturbed to a nearby point, and its stable and unstable manifolds remain close to those of the unperturbed system (*Figure 10.i*).

Melnikov realised that in the case of the perturbed system it was possible to derive an expression for the distance between the stable and unstable manifolds as measured along the normal to Γ_0 and that an estimate for this distance could be computed using the unperturbed solutions. The significance of this lies in the fact that the distance depends on a certain function $M(t_0)$ (now known as the Melnikov function) which, as Melnikov discovered, is an indicator for chaotic motion. It turns out that if $M(t_0)$ contains simple zeros, then the stable and unstable manifolds of

---

[328]Melnikov [1963].

[329]For a review of Melnikov's method see Holmes [1990], where the method is applied to Poincaré's example of a simple pendulum weakly coupled to a linear oscillator. For a more detailed commentary see Guckenheimer and Holmes [1983] and Wiggins [1993].

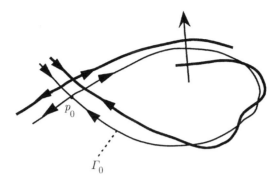

$$p_0$$

$$\Gamma_0$$

FIGURE 10.i

the perturbed system intersect transversely and chaotic motion ensues; but if it is bounded away from zero, then there are no intersections and no chaotic motion.

More generally, Poincaré's qualitative approach involved such a completely new way of looking at dynamical problems, a way so bestrewn with conceptual difficulties, that a period of assimilation was inevitable. This is amply demonstrated by considering Poincaré's method of using a transverse section to reduce the dimension of a dynamical problem in order to simplify the investigation. Although the method now stands as a testimony to Poincaré's remarkable talent for visualising the long-term evolution of a dynamical system, it was some time before its use became widespread. One of the first people to acknowledge the power of the method was Levi-Civita [1901], but this was more than a decade after the publication of Poincaré's memoir. A further twenty years elapsed before Birkhoff began to demonstrate its real potency, and it was only really from then that the idea entered into more common usage.[330]

The mathematical world was unprepared for Poincaré's ideas, but was it better prepared to receive Birkhoff's? Certainly the proof of Poincaré's last theorem meant that interest in Birkhoff and his future research was assured. Nevertheless, while his work was widely acclaimed, the response to it was slow. Although this again can be attributed partly to the novelty of his ideas, there were other indirect factors that inhibited its progress. In global terms there was of course the Great War, which took its inevitable toll on mathematical activity. Then by the time of the publication of Birkhoff's seminal book on dynamical systems [1927], dynamics had to compete in the mathematical arena with not only Einstein's general theory of relativity but also the new ideas of quantum theory. Concomitant with this came a wealth of new mathematics that served to focus interest away from the traditional problems of celestial mechanics and the related dynamics.[331]

---

[330]As, for example, in the work of Thomas Cherry, who in the 1920s did extensive research into the general solutions of equations of dynamical systems. Of note is Cherry's paper [1928] on the periodic solutions of Hamiltonian systems, in which he drew analogies with Poincaré's results.

[331]That is not of course to say that dynamics and these topics were thought to be mutually exclusive. For example, Einstein [1917] showed his familiarity with Poincaré's ideas when he discussed the quantisation conditions for nonseparable but integrable Hamiltonian systems while considering the nature of motion of classical systems with more than one degree of freedom.

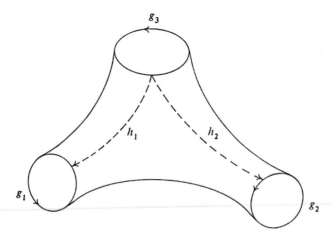

FIGURE 10.ii

However, the thread was not completely cut, and as indicated below, both the work of Marston Morse and that of Kolmogorov, Arnold and Moser bear strong witness to the fundamental importance of Poincaré's work for modern mathematical research into dynamical systems.

10.2. Morse

Marston Morse, a student of Birkhoff's, was deeply influenced by his teacher's interest in Poincaré's topological approach to dynamics. He was born in 1892, the year of the publication of the first volume of Poincaré's *Méthodes Nouvelles*, and became one of the foremost mathematicians of his generation, producing outstanding work in topology and the calculus of variations.

Morse's early papers on geodesics [1920, 1921a], which resulted from his thesis of 1917, drew on results both from Hadamard [1898] and Birkhoff [1912]. In these papers Morse discussed the behaviour of geodesics on surfaces of negative curvature embedded in three-dimensional space, the surfaces being of the type shown in *Figure 10.ii*.[332] These surfaces have boundaries which are closed geodesics g and are assumed to have genus at least equal to two.

Morse, using Hadamard's existence theorems and the fact that if the surface is cut along the geodesic segments h it becomes simply connected, created a symbolic representation of the geodesics lying entirely on the surface. He assigned to each geodesic a sequence of symbols, where each symbol represented a geodesic segment and the whole sequence represented an unending ordered set of geodesic segments, which he termed a *normal set*. Using this representation he obtained results about sets of geodesics and their limit geodesics. Notably he proved the existence of a certain class of geodesics which constituted a set of discontinuous recurrent motions and so resolved Birkhoff's question posed in [1912]. These particular geodesics are now known as *Morse trajectories*.

---

[332]Bott [1980, *915*].

Morse's symbolic representation for the geodesic flow followed naturally from Hadamard's representation of curves on a surface given in [1898]. He took up the idea in more detail at the end of the 1930s, when, with Hedlund, he presented a formalised account of "symbolic dynamics" [1938]. Their work provided the foundations for a powerful new method of dynamical investigation through which dynamical questions could be given an algebraic formulation quite distinct from the classical theories of differential equations.

10.3. KAM theory

Finally, we turn to the question of the convergence of Lindstedt's series. As described in Chapter 7, Poincaré's research had indicated, contrary to what Weierstrass had hoped, that Lindstedt's series were, apart from some exceptional cases, divergent. There was, however, one proviso. Poincaré had made it clear that he had not given a rigorous proof for the cases when the frequencies can be fixed in advance. With the work of Kolmogorov, Arnold and Moser it is now known that in these latter cases the majority of the formal series solutions are in fact convergent, and hence Weierstrass's intuition was after all correct. Their results form the basis for what is now known as Kolmogorov-Arnold-Moser (KAM) theory, which provides methods for integrating perturbed Hamiltonian systems valid for infinite periods of time.

The fundamentals of KAM theory can be understood by considering the autonomous Hamiltonian system with n degrees of freedom

$$\frac{dq}{dt} = \frac{\partial H}{\partial p}, \qquad \frac{dp}{dt} = -\frac{\partial H}{\partial q}, \qquad (p = p_1, \dots, p_n; q = q_1, \dots, q_n)$$

with an analytic Hamiltonian

$$H(p, q) = H_0(p) + \mu H_1(p, q) + \dots,$$

where H is periodic in q of period 2π, and μ is a small parameter.

When the motion is unperturbed and $\mu = 0$,

$$\frac{dp}{dt} = 0, \qquad \frac{dq}{dt} = \frac{\partial H_0}{\partial p} = \omega(p), \qquad (\omega = \omega_1, \dots, \omega_n),$$

the system is integrable and the phase space is foliated by invariant tori $p = constant$. If the frequency ratios ω_i are incommensurable, then the motion is termed quasi-periodic with n frequencies $\omega_1, \dots, \omega_n$, and each trajectory $p(t)$, $q(t)$ is everywhere dense in the torus. The variables p, q are known as action-angle variables, and the unperturbed system is said to be nondegenerate if the motion cannot be described by a smaller number of frequencies than the number of degrees of freedom. In other words the system is nondegenerate providing the Hessian determinant $\left(\dfrac{\partial^2 H_0}{\partial p^2}\right)$ does not vanish identically.

If now the system is slightly perturbed, what happens to the invariant tori? The answer is given in the famous theorem of Kolmogorov [1954]. Clearly, the conditions under which the invariant tori are preserved are precisely the conditions under which solutions to the Hamiltonian equations exist (for fixed frequencies

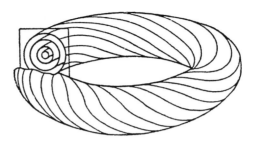

FIGURE 10.iii

independent of the perturbation parameter), and the formal series expansions are convergent.

Kolmogorov's theorem states that if the unperturbed motion is nondegenerate, then, under sufficiently small analytic perturbations of the Hamiltonian, the majority of invariant tori are not destroyed but only shifted slightly in the phase space, and the corresponding motion remains quasi-periodic. Moreover, these invariant tori form a closed nowhere dense set of positive measure whose complement has a measure which is small with respect to μ.

More specifically, the tori which persist under small perturbations are those whose frequencies ω_j are not only incommensurable but also satisfy the inequalities

$$(51) \qquad \left| \sum_{j=1}^{n} k_j \omega_j \right| \geq K \, |k|^{-\nu}, \qquad (\nu = n+1),$$

for all integers k_j with $|k| = \sum |k_j| \geq 1$, and a suitable $K(\omega) > 0$. These relations give a condition under which the small divisors are bounded below in absolute value, and it is known by the theory of Diophantine approximation that for the majority of frequencies these relations are satisfied. These relations are naturally preserved under the perturbations, since Kolmogorov's theorem asserts that the frequencies are independent of the perturbation parameter.

It is necessary to have relations of the type (51) which exclude commensurable frequencies, since in any neighbourhood of an invariant torus of the unperturbed system there exists an invariant torus with frequencies ω_j which are commensurable, and, in general, under a small perturbation such an invariant torus will collapse. These resonant tori consist of periodic solutions, and, as Poincaré had shown when $n = 2$, only a finite number of periodic solutions persist for small values of μ.

Furthermore, when $n = 2$, the two-dimensional invariant tori divide the three-dimensional energy level $H = constant$, and a trajectory originating in the region between two invariant tori remains confined there (*Figure 10.iii*). The measure of the gap between two invariant tori limits the magnitude of the oscillations of the corresponding action variables, and hence these variables remain close to their initial values. When $n > 2$, the n-dimensional tori do not divide the $(2n-1)$-dimensional energy level manifold $H = constant$.

The proof of Kolmogorov's theorem, which was proposed by Kolmogorov [1954] and made rigorous by his student Arnold [1963], was based on Lindstedt's method of the construction of a succession of coordinate changes which progressively annihilate certain terms in the Hamiltonian in increasingly higher order of the parameter. Due to the presence of the small divisors, the convergence of the series satisfying the equations depends on the rate of contraction of the numerators. Kolmogorov's suggested form of proof used Newton's method of approximation, which introduces quadratic convergence, i.e., the error ε_n of the nth approximation is of the order ε_{n-1}^2 for $n = 1, 2, \ldots$, and $\varepsilon < 1$. An important improvement to the theorem was made by Moser [1962], who showed that the requirement of the analyticity of the Hamiltonian can be abandoned and replaced by the condition that several hundred derivatives exist. In the two degree of freedom case he proved that the number of derivatives required is 333.

With the development of KAM theory Weierstrass's question was finally answered in the affirmative. The proof of Kolmogorov's theorem conclusively establishes the existence of convergent series solutions for the n body problem, and, moreover, solutions which are not exceptional in measure-theoretic terms.

Arnold's description of Kolmogorov's theorem in his introduction to the proof encapsulates the spirit of Kolmogorov's achievement:

> *A simple and novel idea, the combination of very classical and essentially modern methods, the solution of a 200 year old problem, a clear geometrical picture and great breadth of outlook ...* .[333]

It is an appropriate tribute to an idea that incorporates much of the rich legacy inherited from Poincaré's great work on the three body problem.

---

[333] Arnold [1963, *9*].

A letter from Gösta Mittag-Leffler
to Sonya Kovalevskaya

Helsingfors, 7.6.1884

. . .

I agree with Weierstrass, if none of the answers on the set question are worthy of the prize, then the medal must be awarded to the mathematician who within recent years has made the best discoveries in higher analysis. But I cannot agree with the view that it will not further the progress of science to propose specific prize questions, in particular if they are stated reasonably. What about the significance for the development of the theory of linear differential equations resulting from the last prize question from the French Academy. This question provided the starting point for Poincaré's work. Furthermore, there does not exist a prize exclusively intended for pure analysis, hardly a prize exclusively intended for pure mathematics, apart from Steiner's prize in Berlin. The shortcoming of the Steiner prize is that it is awarded too often, every year. But we should not award our prize more frequently than every fourth year. Malmsten and the King want the prize jury to be appointed by the King and to consist of

1. The main editor of *Acta Mathematica*
2. A German or Austrian mathematician - = Weierstrass
3. A French or Belgian mathematician - = Hermite
4. An English or American mathematician - = Cayley? or Sylvester
5. A Russian or Italian mathematician - = the first time Brioschi or Tschebychef, the second time Mrs Kovalevskaya.

After each prize giving two of the prize judges should leave the jury and new ones should be appointed by King Oscar as long as he is alive - he must be able to appoint [*substitutes*] for both the leaving members. After King Oscar's death, the three remaining must appoint two new members but always in such a way as to fit the categories mentioned above. Imagine if one had to award a prize to the best mathematical work which has appeared during the last four years. Then the national differences would certainly show up and the different views on what constitutes the essential substance of mathematics would be clearly expressed. Cayley and Brioschi might want to make the award to God knows which master of calculating and Tschebychef might opt for God knows what odd ideas. It is quite a different matter when one has to judge answers to a specific question. In this case one is forced to stick to much more objective criteria.

And finally one more reason. King Oscar is convinced that we should only announce specific prize questions and I doubt that it will be possible to change his mind unless I propose that the prize should be used to honour Swedish or Norwegian papers - like the anatomic prize you mentioned which is probably restricted to Russian works - but I do not want this at all. The competition would not get the international scientific reputation which I had imagined. I think that honouring only the works published in Acta would be more in the interests of the journal than in the interests of science. And this I do not want either. The interests of science must come first. If later I can do something for Acta at the same time then that is another matter and I will do it with all my heart.

Announcement of the Oscar Competition

Nature 30.7.1885

THE HIGHER MATHEMATICS

Prof. G. Mittag-Leffler, principal editor of the *Acta Mathematica*, forwards us the following communication, which will shortly appear in that journal:-

His Majesty Oscar II, wishing to give a fresh proof of his interest in the advancement of mathematical science, an interest already manifested by his graciously encouraging the publication of the journal *Acta Mathematica*, which is placed under his august protection, has resolved to award a prize, on January 21, 1889, the sixtieth anniversary of his birthday, to an important discovery in the field of higher mathematical analysis. This prize will consist of a gold medal of the eighteenth size bearing his Majesty's image and having a value of a thousand francs, together with a sum of two thousand five hundred crowns (1 crown = about 1 franc 4- centimes).

His Majesty has been pleased to entrust the task of carrying out his intentions to a commission of three members, Mr. Karl Weierstrass in Berlin, Mr. Charles Hermite in Paris, and the chief editor of this journal, Mr. Gösta Mittag-Leffler in Stockholm. The commissioners having presented a report on their work to his Majesty, he has graciously signified his approval of the following final propositions of theirs.

Having taken into consideration the questions which from different points of view equally engage the attention of analysts, and the solution of which would be of the greatest interest for the progress of science, the commission respectfully proposes to his Majesty to award the prize to the best memoir on one of the following subjects:-

(1) A system being given of a number whatever of particles attracting one another mutually according to Newton's law, it is proposed, on the assumption that there never takes place an impact of two particles to expand the coordinates of each particle in a series proceeding according to some known functions of time and converging uniformly for any space of time.

It seems that this problem, the solution of which will considerably enlarge our knowledge with regard to the system of the universe, might be solved by means of the analytical resources at our present disposition; this may at least be fairly supposed, because shortly before his death Lejeune-Dirichlet communicated to a friend of his, a mathematician, that he had discovered a method of integrating the differential equations of mechanics, and that he had succeeded, by applying this method, to demonstrate the stability of our planetary system in an absolutely strict manner. Unfortunately we know nothing about this method except that the starting point for its discovery seems to have been the theory of infinitely small oscillations.[334] It may, however, be supposed almost with certainty that this method was not based on long and complicated calculations but on the development of a simple fundamental idea, which one may reasonably hope to find again by means of earnest and persevering study.

However, in case no one should succeed in solving the proposed problem within the period of the competition, the prize might be awarded to a work in which some other problem of mechanics is treated in the indicated manner and completely solved.

(2) Mr. Fuchs has demonstrated in several of his memoirs[335] that there exist uniform functions of two variables which, by their mode of generation, are connected with the ultra-elliptical functions, but are more general than these, and which would probably acquire great importance for analysis, if their theory were further developed.

It is proposed to obtain in an explicit form those functions whose existence has been proved by Mr. Fuchs, in a sufficiently general case, so as to allow of an insight into and study of their most essential properties.

(3) A study of the functions defined by a sufficiently general differential equation of the first order, the first member of which is a rational integral function with respect to the variable, the function, and its first differential coefficient.

Mr. Briot and Mr. Bouquet have opened the way for such a study by their memoir on this subject (*Journal* de l'École polytechnique, cahier 36, pp. 133-198). But mathematicians acquainted with the results attained by these authors know also that their work has not by any means exhausted the difficult and important subject which they have first treated. It seems probable that, if fresh inquiries were to be undertaken in the same direction, they might lead to theorems of high interest for analysis.

(4) It is well known how much light has been thrown on the general theory of algebraic equations by the study of the special functions to which the division of the circle into equal parts and the division of the argument of the elliptic functions by a whole number lead up. That remarkable transcendant which is obtained by expressing the module of an elliptic function by the quotient of the periods leads likewise to the modular equations, that have been the origin of entirely new notions and highly important results, as the solution of equations in the fifth degree. But

---

[334]See p. 35 of the Panegyric on Lejeune-Dirichlet by Kummer, "Abhandlungen der K. Akademie der Wissenschaften zu Berlin," 1860.

[335]These memoirs are to be found in (1) "Nachrichten von der K. Gesellschaft der Wissenschaften zu Göttingen," February, 1880, p. 170; (2) Borchardt's "Journal," Bd. 89, p. 251 (a translation of this memoir is to be found in the "Bulletin" of Mr. Darboux, 2me série, t.iv); (3) "Nachrichten von der K. Gesellschaft der Wissenschaften zu Göttingen," June, 1880, p. 445 (translated into French in the "bulletin" of Mr. Darboux, 2me série, t.iv); (4) Borchardt's "journal," Bd. 90, p. 71 (also in the "Bulletin of Mr. Darboux, 2me série, t.iv); (5) "Abhandlungen der K. Gesellschaft der Wissenschaften zu Göttingen," 1881 ("Bulletin" of Mr. Darboux, t.v); (6) "Sitzungsberichte der K. Akademie der Wissenschaften zu Berlin" 1883, i, p. 507; (7) The memoir of Mr. Fuchs published in Borchardt's "Journal," Bd. 76, p. 177, has also some bearings on the memoirs quoted.

this transcendant is but the first term, a particular case and that the simplest one of an infinite series of new functions introduced into science by Mr. Poincaré under the name of "fonctions fuchsiennes," and *successfully* applied by him to the integration of linear differential equations of any order. These functions, which accordingly have a *rôle* of manifest importance in analysis, have not as yet been considered from an algebraical point of view as the transcendant of the theory of elliptic functions of which they are the generalisation.

It is proposed to fill up this gap and to arrive at new equations analogous to the modular equations by studying, though it were only in a particular case, the formation and properties of the algebraic relations that connect two "fonctions fuchsiennes" when they have a group in common.

In case none of the memoirs tendered for competition on any of the subjects proposed above should be deemed worthy of the prize, this may be adjudged to a memoir sent in for competition that contains a complete solution of an important question of the theory of functions other than those proposed by the Commission.

The memoirs offered for competition should be furnished with an epigraph and, besides, with the author's name and place of residence in a sealed cover, and directed to the chief editor of the *Acta Mathematica* before June 1, 1888.

The memoir to which his Majesty shall be pleased to award the prize as well as that or those memoirs which may be considered by the Commission worthy of an honorary mention, will be inserted in the *Acta Mathematica*, nor can any of them be previously published.

The memoirs may be written in any language that the author chooses, but as the members of the Commission belong to three different nations the author ought to subjoin a French translation to his original memoir, in case it is not written in French. If such a translation is not subjoined the author must allow the Commission to have one made for their own use.

THE EDITORS OF ACTA MATHEMATICA

Entries received in the Oscar Competition

The following pages contain the announcement of the entries received in the Oscar Competition (*Acta*, **11** (1888), 401–402). The titles of the entries received for the competition are listed in the order in which they were received.

Prix Oscar II.

Mémoires présentés au concours.

Le concours pour le prix fondé par S. M. le roi OSCAR II a été clos le 1^r juin de cette année. Nous mentionnons ci-après et dans l'ordre où ils sont parvenus, les mémoires destinés au concours qui ont été adressés au Rédacteur en chef de ce journal, à Stockholm:

1. *Mémoire sur l'équation trinôme de degré impair $x^m \pm x = r$.*

Épigraphe: Les trois nombres harmoniques élémentaires sont 2, 3 et 5.

2. *Nuova Teoria dei Massimi e Minimi degli Integrali definiti.*

Épigraphe: Opinionum commenta delet dies; naturæ judicia confirmat.

(Cic. Nat. D.)

3. *Allgemeine Entwicklung der Functionen.*

Épigraphe: Sich selbst zu loben ist ein Fehler,
Doch jeder thut's, der etwas Gutes thut.

(Westöstlicher Divan von Göthe.)

L'auteur y a joint une traduction française:

Développement général des fonctions

avec l'épigraphe: Tu ne fais pas bien en te louant toi-même
Mais tu te loues toi-même en faisant bien.

(D'après Goethe.)

4. *Les Fonctions Pseudo- et Hyper-Bernoulliennes et leurs premières applications.* — Contribution élémentaire à l'intégration des équations différentielles.

Épigraphe: Venient qui sine offensa, sine gratia, judicent.

(Senèque.)

5. *Über die Bewegungen in einem System von Massepunkten mit Kräften der Form* $-\frac{1}{r^2}$.

Épigraphe: Ἁπλοῦς ὁ λόγος τῆς ἀληθείας ἔφυ.

(Euripides.)

6. *Intégration des équations simultanées aux dérivées partielles du premier ordre d'un nombre quelconque de fonctions de plusieurs variables indépendantes.*

Épigraphe: Accipe jussis
carmina cepta tuis.

7. *Über die Integration der Differentialgleichungen, welche die Bewegungen eines Systems von Puncten bestimmen.*

Épigraphe: Nur schrittweise gelangt man zum Ziel.

Avec une traduction française, intitulée:

Sur l'intégration des équations différentielles qui déterminent les mouvements d'un système de points matériels,

et portant l'épigraphe: Pour parvenir au sommet, il faut marcher pas à pas.

8. *Sur les intégrales de fonctions à multiplicateurs et leur application au développement des fonctions abéliennes en séries trigonométriques.*

Épigraphe: Nous devons l'unique science
Que l'homme puisse conquérir
Aux chercheurs dont la patience
En a laissé les fruits mûrir.
(Sully-Prudhomme, Le Bonheur.)

Avec un Supplément.

9. *Sur le Problème des trois Corps et les Équations de la Dynamique.*

Épigraphe: Nunquam præscriptos transibunt sidera fines.

10. *Sur le Problème des trois Corps.*

Épigraphe: -- — -- — -- — Coelumque tueri
Jussit et erectos ad sidera tollere vultus.
(Ovide.)

11. *Über die Bewegung der Himmelskörper im widerstehenden Mittel.*

Épigraphe: Per aspera ad astra.

12. *Recherches sur la formule sommatoire d'Euler.*

Épigraphe: Utinam ne nimis erraverim.

Juin 1888.

MITTAG-LEFFLER.

Report of the Prize Commission

Poincaré *Œuvres* **XI**, *286-289*

French translation sent to Poincaré after the announcement of the competition result.

Traduction

Procès-verbal dressé par devant S. M. le Roi au palais de Stockholm, le 20 Janvier 1889, en présence de S. Exc. M. le Comte Ehrensvärd, Ministre des Affaires Etrangères, M. G. Wennerberg, Ministre des Cultes et de l'Instruction Publique, M. R. O. Schjött, Ministre Norvegien et de M. G. Mittag-Leffler, professeur à l'université de Stockholm.

§1. La commission, nommeé par S. M. le Roi, en date du 25 Novembre 1884, pour examiner des mémoires, ayant concouru pour le prix en mathématiques offert par Sa Majesté, et composé de M. Carl Weierstrass, professeur à l'université de Berlin, M. Charles Hermite, professeur à la Sorbonne à Paris, et M. Gösta Mittag-Leffler, professeur à l'université de Stockholm, ayant terminé ses travaux, le rapport de la commission fut soumis au Roi.

Il ressort de ce rapport que la commission a été de l'opinion unanime, que le mémoire qui est intitulé *Sur le problème des trois corps et les équations de la dynamique* avec la devise "Nunquam præscriptos transibunt sidera fines", est l'œuvre profonde et originale d'un génie mathématique dont la place est marqué parmi les grands géomêtres du siècle. Les plus importantes et les plus difficiles questions, comme la stabilité du système du monde, l'expression analytique des coordonnées des planètes par des series de sinus et de cosinus des multiples du temps, puis l'étude on ne peut plus remarquable, des mouvements asymptotiques, la découverte de formes de mouvement où les distances des corps restant comprises entre des limites fixes, on ne peut cependant exprimer leurs coordoneés par des séries trigonométriques, d'autres sujets encore que nous n'indiquons point, sont traités par des méthodes qui ouvrent, il n'est que juste de le dire, une époque nouvelle dans la mécanique céleste. Les notions analytiques inconnues de Lagrange et de Laplace, qui n'ont été acquises que de notre temps, ont un rôle essentiel dans ces questions si difficiles où le talent de l'auteur se montre dans tout son éclat. Une fois de plus se trouve ainsi confirmé cette observation que les plus grands progrés en astronomie, en physique et les découvertes qui étendent le domaine des mathématiques abstraites, se produisent simultanément, comme si elles étaient appelées à se seconder en concourant à un même but, et que la commission de même a été unanime dans l'opinion, que l'auteur du mémoire qui porte pour titre *Sur les intègrales des fonctions à multiplicateurs et leur application au développement des fonctions abéliennes en séries trigonometriques,* et a pour devise

> "Nous devons l'unique science
> Que l'homme puisse conquérir
> Aux chercheurs dont la patience
> En a laissé les fruits mûrir." [336]

a montré un talent mathématique de premier ordre, et que son mémoire est extrèmement digne de l'attention des géomètres.

§2. S. M. le Roi daigna décerner le prix offert par Sa Majesté et composé d'une médaille sen or evaluée à environs 1,000 francs ainsi que la somme 2,500 couronnes à l'auteur de mémoire muni de l'épigraphe "Nunquam præscriptos transibunt sidera fines" et un exemplaire de la médaille à l'effigie de Sa Majesté et portant l'inscription "in sui memoriam" à l'auteur de mémoire portant l'épigraphe:

> "Nous devons l'unique science ... ."

§3. S. M. le Roi ayant en suite ouvert les bulletins accompagnant les dit mémoires, il a été constaté que le bulletin à l'épigraphe: "Nunquam præscriptos transibunt sidera fines" portait le nom "M. H. Poincaré, Paris", et celui à l'épigraphe:

> "Nous devons l'unique science ..."

le nom de "Paul Appell, Paris".

Ainsi passé: Au Château de Stockholm le 20 Janvier 1889.

> Oscar
> Alb. Ehrensvärd G. Wennerberg
> P. O. Schjött G. Mittag-Leffler

> Otto Printzsköld

---

[336]Sully-Prudhomme, Le Bonheur.

Title Pages and Tables of Contents

The following pages contain the title page and the table of contents for [P1] and [P2], the unpublished and the published versions of Poincaré's memoir on the three body problem.

5.1. Poincaré's Unpublished Memoir

SUR LE

PROBLÈME DES TROIS CORPS

ET LES

ÉQUATIONS DE LA DYNAMIQUE

PAR

H. POINCARÉ

à PARIS.

———

MÉMOIRE COURONNÉ

DU PRIX DE S. M. LE ROI OSCAR II

LE 21 JANVIER 1889.

———

AVEC DES NOTES

PAR L'AUTEUR.

TABLE DES MATIÈRES.

Notes.

5.2. Poincaré's Published Memoir

SUR LE

PROBLÈME DES TROIS CORPS

ET LES

ÉQUATIONS DE LA DYNAMIQUE

PAR

H. POINCARÉ
à PARIS.

———

MÉMOIRE COURONNÉ
DU PRIX DE S. M. LE ROI OSCAR II
LE 21 JANVIER 1889.

TABLE DES MATIÈRES.

APPENDIX 6

Theorems in [P1] not included in [P2]

In the chapter on invariant integrals in [P1] Poincaré included two extensions to Theorem III, as well as a further theorem, Theorem IV. These theorems are not included in [P2] because either they became redundant as a result of the discovery of the error or they had no particular relevance to any other part of the memoir.

First Extension to Theorem III. *Suppose that $A_n B_n$ coincides partly with $A_0 B_0$ and partly with the extension of $A_0 B_0$ such that $A_0 B_0$ is an nth order invariant curve. Suppose also that the distance between A_0 and B_p is a small quantity of qth order, where p is prime to n, then the distance from A_0 to its nth consequent A_n is a very small quantity of qth order.*

This is the case described by *Figure 6.i* where n is taken to be 5.

Combining this generalisation with the Corollary, Poincaré deduced a further generalisation in which he gave the conditions under which he claimed that the set of curves formed by $A_0 B_0$ and its successive consequents would form a "closed" invariant curve of first order. By invoking the Corollary he again reiterated the error he had made earlier.

Poincaré prefaced the second extension to Theorem III [P1] by saying that he did not expect to use it in what followed, although he later cited it twice, both references becoming redundant in the revision.

Second Extension to Theorem III. *A curve without being rigorously invariant may be invariant up to a very small quantity of pth order. If the distance*

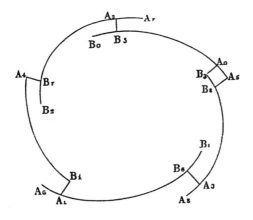

Figure 6.i

between a curve C, which is not rigorously invariant, and an arbitrary point of its nth consequent is a very small quantity of pth order, then such a curve is called nth order semi-invariant up to very small quantities of pth order. If a semi-invariant curve is quasi-closed such that the distance between the points of closure A and B is very small of qth order, the distance from the point A to its nth consequent A_n will be very small of order at least q providing $2q < p$, and of order $p - q$ providing $2q > p > q$.

THEOREM IV. *Consider a transverse section S which is simply connected. Let a point on S be determined by a particular system of coordinates (to be defined) which is analogous to polar coordinates. Let O be an arbitrary point on S at which infinitely many branches of a curve meet, in the same way that radius vectors meet at the pole in polar coordinates. Suppose that O is the only common point of any two branches of the curve and that an arbitrary branch is defined by the angle q between its tangent at O and a fixed line passing through O.*

Consider a second system of closed concentric curves containing the point O. Furthermore, suppose that any curve of the second system has one and only one point in common with any curve of the first system. Consider a fixed branch of the first system B_0 and let P be the point where it cuts a moving curve of the second system. Let r be the length of the arc of the curve B_0 between O and P. The moving curve can then be defined by r. Finally suppose that through an arbitrary point P of S there passes one and only one branch of the first system. The coordinates r and q can then be used to define the position of P on S.

Let a be a simply connected area of S limited by a closed curve k. Let a_n be the nth consequent limited by the closed curve k_n. If the two areas a and a_n have a part in common and O belongs to this communal part, if the points of k have same coordinate q as their n consequents, if the curve k meets each of the branches of the first system at one point (such that when one crosses the closed curve k, q varies between 0 and 2π), if there is a positive invariant integral, then two at least of the points k coincide with their nth consequents.

Poincaré also gave a second, more succinct, statement of the theorem which did not involve the coordinate system defined above. Let k be a closed curve on a simply connected transverse section S with nth consequent k_n. If each of the points of k can be joined to its nth consequent by arcs of curves on S in such a way that no two of these arcs have a point in common, and, moreover, there is a positive invariant integral, two at least of the points of k will coincide with their consequents.

There appears to be no clear reason why Poincaré included this Theorem in [P1], as he made no use of it there. His only reference to it was an expression of regret that he did not have the opportunity to show how it could be applied in the study of the spatial distribution of closed trajectories. It may have been because he was writing to a deadline that he decided it was altogether easier to keep it in his competition entry, or it may have been simply that he was not good at organising his material. Alternatively, it may have been because he decided that having established the result it made sense to publish it so that it was available should he ever need it, although it does not appear that he ever made use of it.

References

Items listed include all published works referred to in the text and some others found useful in its preparation.

Abbreviations

Acta   Acta Mathematica
BAAS   British Association for the Advancement of Science
Cahiers   Cahiers de Séminaire d'Histoire des Mathématiques
Comptes Rendus   Comptes Rendus Hebdomadaires des Séances de l'Académie des Sciences
I M-L   Institut Mittag-Leffler

R. C. Archibald

[1936] Unpublished Letters of James Joseph Sylvester and other new Information concerning his Life and Work, *Osiris* **1**, 85-154.

V. I. Arnold

[1963] Proof of a theorem of A. N. Kolmogorov on the invariance of quasi-periodic motions under small perturbations of the Hamiltonian, *Russian Mathematical Surveys* **18**, No. 5, 9-36.

[1963a] Small denominators and problems of stability of motion in classical and celestial mechanics, *Russian Mathematical Surveys* **18**, No. 6, 85-192.

[1988] *Dynamical Systems III*, Springer-Verlag, Berlin.

H. F. Baker

[1914] Henri Poincaré, *Proceedings of the Royal Society A* **91**, vi-xvi.

[1916] On certain linear differential equations of astronomical interest, *Philosophical Transactions of the Royal Society A* **216**, 129-186.

I. Bendixson

[1901] Sur les courbes définies par des équations différentielles, *Acta* **24**, 1-88.

K-R. Biermann

[1988] *Die Mathematik und ihre Dozenten an der Berliner Universität 1810-1933*, 2nd edition, Akademie-Verlag, Berlin.

G. Birkhoff

[1950] *George David Birkhoff Collected Mathematical Papers*, 3 volumes, American Mathematical Society, New York. Republished Dover, New York, 1968.

[1912] Quelques théoremes sur le mouvement des systèmes dynamiques, *Bulletin de la Société Mathématique de France* **40**, 305-323 = *Collected Mathematical Papers* **I**, 654-672.

[1913] Proof of Poincaré's geometric theorem, *Transactions of the American Mathematical Society* **14**, 14-22 = *Collected Mathematical Papers* **I**, 673-681.

[1915] The restricted problem of three bodies, *Rendiconti del Circolo Matematicodi Palermo* **39**, 265-334 = *Collected Mathematical Papers* **I**, 682-751.

[1917] Dynamical systems with two degrees of freedom, *Transactions of the American Mathematical Society* **18**, 199-300 = *Collected Mathematical Papers* **II**, 1-102.

[1920] Recent advances in dynamics, *Science* **51** No. 1307, January 16, 51-55 = *Collected Mathematical Papers* **II**, 106-110.

[1920a] Surface transformations and their dynamical applications, *Acta* **43**, 1-119 = *Collected Mathematical Papers* **II**, 111-229.

[1922] Celestial Mechanics, *Bulletin of the National Research Council* **4** (Part 1), 1-22 = *Collected Mathematical Papers* **II**, 252-266.

[1925] An extension of Poincaré's last geometric theorem, *Acta* **47**, 297-311 = *Collected Mathematical Papers* **II**, 230-251.

[1927] *Dynamical Systems*, American Mathematical Society, Providence, RI. Reprinted 1966 with additional references by Jurgen Moser.

[1927a] Stability and the equations of dynamics, *American Journal of Mathematics* **49**, 1-38 = *Collected Mathematical Papers* **II**, 295-332.

[1927b] On the periodic motions of dynamical systems, *Acta* **50**, 459-379 = *Collected Mathematical Papers* **II**, 333-353.

[1928] A remark on the dynamical role of Poincaré's last geometric theorem, *Acta Litterarum ac Scientiarum* **4**, 6-11 = *Collected Mathematical Papers* **II**, 354-359.

[1931] Une généralisation à *n* dimensions du dernier théorème de géométrie de Poincaré, *Comptes Rendus* **192**, 196-198 = *Collected Mathematical Papers* **II**, 395-397.

[1931a] Proof of the ergodic theorem, *Proceedings of the National Academy of Science, USA* **17**, 656-660 = *Collected Mathematical Papers* **II**, 404-408.

[1935] Sur le problème restreint des trois corps (Premier mémoire), *Annali della R. Scuola Normale Superiore Pisa (2)* **4**, 267-306 = *Collected Mathematical Papers* **II**, 466-505.

[1935a] Nouvelles recherches sur les systèmes dynamiques, *Memoriae Pontifical Academia della Scienze Novi Lyncaei (3)* **1**, 85-216 = *Collected Mathematical Papers* **II**, 530-661.

[1936] Sur le problème restreint des trois corps (Second mémoire), *Annali della R. Scuola Normale Superiore Pisa (2)* **5**, 1-42 = *Collected Mathematical Papers* **II**, 668-709.

[1942] What is the ergodic theorem? *American Mathematical Monthly* **49**, 222-226 = *Collected Mathematical Papers* **II**, 713-717.

G. Bisconcini

[1906] Sur le problème des trois corps. Trajectoires le long desquelles deux au moins des trois corps se choquent. Conditions qui entraînent un choc, *Acta* **30**, 49-92.

K. Bohlin

[1887] Über die Bedeutung des Princips der lebendigen Kraft für die Frage von der Stabilität dynamischer Systeme, *Acta* **10**, 109-130.

[1888] Ueber eine neue Annäherungsmethode in der Störungstheorie, *Bihang till Svenska Vetenskap Akademiens handlingar* **14**, No. 5; Zur Frage der Convergenz der Reihenentwickelungen in der Störungstheorie, *Astronomische Nachrichten* **121**, 17-24.

L. Boltzmann

[1871] Ueber die Druckkräfte, welche auf Ringe wirksam sind, die in bewegte Flüssigkeit tauchen, *Journal für die Reine und Angewandte Mathematik* **73**, 111-134.

R. Bott

[1980] Marston Morse and his mathematical works, *Bulletin of the American Mathematical Society* **3**, 907-944.

O. R. M. Brendel

[1889] On the problem of three bodies according to Gyldén's theory, *The Observatory* (1989) 399-403.

C. Briot and J. Bouquet

[1854] Recherches sur les propriétés des fonctions définies par des équations différentielles, *Comptes Rendus* **39**, 368-371.

[1856] Étude des fonctions d'une variable imaginaire; Recherches sur les propriétés des fonctions définies par des équations différentielles, *Journal de l'École Polytechnique* **21**, 85-132; 133-198.

D. Brouwer and G. M. Clemence

[1961] *Methods of Celestial Mechanics*, Academic Press, New York.

E. W. Brown

[1892] Poincaré's Mécanique Céleste, *Bulletin of the New York Mathematical Society* **I**, 206-214.

[1896] *An Introductory Treatise on the Lunar Theory*, Cambridge University Press, London and New York. Reprinted by Dover, New York, 1960.

[1916] Biographical Memoir of George William Hill, *Biographical Memoirs, National Academy of Sciences* **8**, 275-306.

[1916a] The scientific work of Sir George Darwin, *G. H. Darwin, Scientific Papers* **V**, Cambridge University Press, xxxiv-lv.

H. Bruns

[1887] Über die Integrale des Vielkörper-Problems, *Acta* **11**, 25-96.

S. G. Brush

[1966] *Kinetic Theory*, 2 volumes, Pergamon Press, Oxford.

[1980] Poincaré and cosmic evolution, *Physics Today* **33**, 42-49.

G. Cantor

[1872] Über die Ausdehnung eines Satzes aus der Theorie der trigonometrischen Reihen, *Mathematische Annalen* **5**, 123-132.

C. Carathéodory

[1919] Über der Wiederkehrsatz von Poincaré, *Sitzungsberichte der Preussischen Akademie der Wissenschaften*, 580-584.

E. Cartan

[1922] *Leçons sur les invariants intégraux*, A. Hermann, Paris.

M. L. Cartwright

[1965] Jacques Hadamard, *Journal of the London Mathematical Society* **40**, 722-748.

A.-L. Cauchy

[1842] Mémoire sur un théorème fondamental dans le calcul intégrale, *Comptes Rendus* **14**, 1020–1026.

A. Cayley

[1862] Report on the progress of the solution of certain special problems in dynamics, *BAAS Report 1862*, 184-252 = *The Collected Mathematical Papers of Arthur Cayley* **IV**, 513-593.

L. Cesari

[1959] *Asymptotic behaviour and stability problems in ordinary differential equations*, Springer-Verlag, Berlin.

R. P. Cesco

[1961] Some Theorems and Results in the Three Body Problem, *Proceedings of the International Meeting on Problems of Astrometry and Celestial Mechanics, La Plata, Argentina*, 81-92.

J. Chazy

[1920] Sur les singularities impossible du problème des n corps, *Comptes Rendus* **170**, 575-577.

[1922] Sur l'allure du mouvement dans le problème des trois corps, *Annales Scientifiques de l'École Normale Superieure* **39** (Series 3), 29-130.

[1952] L'intégration du problème des trois corps par Sundman, et ses conséquences, *Bulletin Astronomique* **16**, 175-190.

T. M. Cherry

[1924] On Poincaré's theorem of "The non-existence of uniform integrals of dynamical equations", *Proceedings of the Cambridge Philosophical Society* **22**, 287-294.

[1928] On periodic solutions of Hamiltonian systems of differential equations, *Philosophical Transactions of the Royal Society A* **227**, 137-221.

A. C. Clairaut

[1747] Du Système du Monde dans les Principes de la Gravitation Universelle, *Mémoires de l'Académie royale des Sciences de Paris*, 329-364.

[1749] Du Système du Monde dans les Principes de la Gravitation Universelle, *Mémoires de l'Académie royale des Sciences de Paris*, 549.

R. Cooke

[1984] *The Mathematics of Sonya Kovalevskaya*, Springer-Verlag, New York.

G. Darboux

[1914] Éloge historique d'Henri Poincaré lu dans la séance publique annuelle du 15 décembre 1913, *Revue Scientifique* 25 July 1914, 97-110 = Poincaré *Œuvres* **II**, vii-lxxi.

Sir Francis Darwin

[1916] Memoir of Sir George Darwin, *G. H. Darwin, Scientific Papers* **V**, ix-xxxii.

Sir George Darwin

[1911–1916] *Scientific Papers by Sir George Howard Darwin*, 5 volumes, Cambridge University Press.

[1887] On figures of equilibrium of rotating masses of fluid, *Philosophical Transactions of the Royal Society* **178A**, 379-428 = *Scientific Papers* **III**, 135-185.

[1897] Periodic Orbits *Acta* **21**, 99-242 and (with omission of certain tables of results) *Mathematischen Annalen* **51**, 523-583 = *Scientific Papers* **IV**, 1-113.

[1900] Presentation of the medal of the Royal Astronomical Society to M. Henri Poincaré, *Monthly Notices of the Royal Astronomical Society* **60**, 406-415 = *Scientific Papers* **IV**, 510-519.

[1909] On certain families of periodic orbits, *Monthly Notices of the Royal Astronomical Society* **70**, 108-143 = *Scientific Papers* **IV**, 140-181.

J. W. Dauben

[1979] *Georg Cantor. His mathematics and philosophy of the infinite*, Harvard University Press, Cambridge, Massachusetts.

T. de Donder

[1901] Étude sur les invariants integraux, *Rendiconti del Circolo Matematico di Palermo* **15**, 66-131.

C. E. Delaunay

[1846] Nouvelle théorie analytique du mouvement de la lune, *Comptes Rendus* **23**, 968-970.

[1860] Théorie du mouvement de la lune I, *Mémoire de l'Academie des Sciences* **28**, 1-883.

[1867] Théorie du mouvement de la lune II, *Mémoire de l'Academie des Sciences* **29**, 1-931.

L. Dell'Aglio and G. Israel

[1989] La théorie de la stabilité et l'analyse qualitative des équations differentielles ordinaires dans les mathématiques Italiennes: le point de vue de Tullio Levi-Civita, *Cahiers* **10**, 283-321.

F. N. Diacu

[1993] Painlevé's conjecture, *The Mathematical Intelligencer* **15** (2), 6-12.

Y. Domar

[1982] On the foundation of Acta Mathematica, *Acta* **148**, 3-8.

G. Duffing

[1918] *Erzwungene Schwingungen bei veränderlicher Eigenfrequenz und ihre technische Bedeutung*, Braunschweig.

P. Duhem

[1962] *The Aim and Structure of Physical Theory*, Atheneum, New York. Translated by Philip P. Weiner from *La Théorie Physique: Son Objet, Sa Structure*, Marcel Rivière & Cie, Paris, 2nd edition, 1914. First edition published 1906.

A. Einstein

[1917] On the quantisation condition of Sommerfeld and Epstein, *Deutsche Physikalische Gesellschaft Berlin Verhardlungen* **19**, No. 9/10, translated by C. Jaffé, *Joint Institute for Laboratory Astrophysics Report* **116** (1980), University of Colorado.

I. Ekeland

[1988] *Mathematics and the Unexpected*, The University of Chicago Press, Chicago.

L. Euler

[1744] *Theoria motuum planetarum et cometarum = Opera* (2)**28**, 105-251.

[1748] *Recherches due la questions des inégalities du mouvement de Saturne et de Jupiter, sujet propose pour le prix de l'année 1748 pa l'Academie Royale des Sciences de Paris = Opera* (2) **25**, 45-157.

[1753] *Theoria motus lunae = Opera* (2) **23**, 64-336.

[1772] *Theoria motuum lunae, nova methodo pertractata = Opera* (2) **22**, 1-411.

E. Fürstenau

[1860] *Darstellung der reellen Wurzeln algebraischer Gleichungen durch Determinanten der Coeffizienten*, Marburg.

A. Gautier

[1817] *Essai historique sur le problème des trois corps ou dissertation sur la théorie des mouvements de la lune et des planètes*, Courcier, Paris.

J. Gerver

[1991] The existence of pseudocollisions in the plane, *Journal of Differential Equations* **89**, 1-68.

G. E. O. Giacaglia

[1972] *Perturbation Methods in Non-Linear Systems*, George Allen & Unwin Ltd., London.

C. Gilain

[1991] La théorie qualitative de Poincaré et le problème de l'intégration des équations différentielles, *La France Mathématique,* ed. H. Gispert, Cahiers d'histoire et de philosophie des sciences **34**, 215-242.

D. L. Goroff

[1993] Henri Poincaré and the birth of chaos theory: An introduction to the English translation of Les Méthodes Nouvelles de la Mécanique Céleste, New Methods of Celestial Mechanics, American Institute of Physics Press, History of Modern Physics and Astronomy **13**, I1–I107.

I. Grattan-Guinness

[1971] Materials for the history of mathematics in the Institut Mittag-Leffler, *Isis* **62**, 363-374.

J. J. Gray

[1984] Fuchs and the theory of differential equations, *Bulletin of the American Mathematical Society* **10** (1), 1-26.

[1992] Poincaré, topological dynamics, and the stability of the solar system, *An investigation of difficult things. Essays on Newton and the History of Exact Sciences* (ed. P. M. Harman and E. E. Shapiro), Cambridge University Press, 503-524.

J. Guckenheimer and P. Holmes

[1983] *Nonlinear Oscillations, Dynamical Systems, and Bifurcations of Vector Fields*, Springer-Verlag, New York.

H. Gyldén

[1887] Untersuchungen über die Convergenz der Reigen, welche zur Darstellung der Coordinaten der Planeten angewendet werden, *Acta* **9**, 185-294.

[1891] Nouvelles recherches sur les séries employées dans les théories des planètes, *Acta* **15**, 65-189.

[1893] Nouvelles recherches sur les séries employées dans les théories des planètes, *Acta* **17**, 1-168.

J. Hadamard

[1968] *Œuvres de Jacques Hadamard*, 4 volumes, Editions du Centre National de la Recherche Scientifique, Paris.

[1897] Sur certaines propriétés des trajectoires en dynamique, *Journal de Mathématiques* (5) **3**, 331-387 = *Œuvres* **4**, 1749-1805.

[1898] Les surfaces à courbures opposées et leurs lignes géodesiques, *Journal de Mathématiques* (5) **4** , 27-73 = *Œuvres* **2**, 729-780.

[1901] *Notice sur les traveaux scientifiques de Jacques Hadamard*, Gauthier Villars, Paris. Reprinted Hermann, Paris, 1912.

[1912] L'œuvre d'Henri Poincaré, *Revue de Metaphysique et de Morale* **21**, 617-658.

[1913] Henri Poincaré et le problème des trois corps, *Revue du Mois* **16**, 385-419 = Henri Poincaré: L'œuvre scientifique. L'œuvre philosophique, *Nouvelle Collection Scientifique*, Paris, 1914, 51-114 = *Œuvres* **4**, 2007-2041.

[1915] Sur un memoire de Sundman, *Bulletin des sciences mathématiques* **39**, 249-264 = *Œuvres* **4**, 1897-1912.

[1921] L'œuvre mathématique de H. Poincaré, *Acta* **38**, 203-287= *Œuvres* **4**, 1921-2005 = *Œuvres de Henri Poincaré* **11**, 152-242.

[1922] The early scientific work of Henri Poincaré, *Rice Institute Pamphlet* **9**, 111-183.

[1933] The later scientific work of Henri Poincaré, *Rice Institute Pamphlet* **20**, 1-86.

Y. Hagihara

[1970-1976] *Celestial Mechanics* **1-2**, MIT Press, Cambridge, Massachusetts, 1970-1972; **3-5**, Japan Society for the Promotion of Science, Tokyo, 1974-1976.

M. Hamy

[1892] Les Méthodes Nouvelles de la Mécanique céleste I, *Revue Générale des Sciences pures et appliquées*, 649.

[1896] Les Méthodes Nouvelles de la Mécanique céleste II, *Revue Générale des Sciences pures et appliquées*, 39-40.

[1900] Les Méthodes Nouvelles de la Mécanique céleste III, *Revue Générale des Sciences pures et appliquées*, 254-255.

T. Hawkins

[1992] Jacobi and the birth of Lie's theory of groups, *Archive for History of Exact Sciences* **42** (3), 187-278.

M. Hénon

[1965] Exploration numérique du problème restreint, *Annales d'Astrophysique* **28**, 499-511, 992-1007.

C. Hermite

[1877] Sur quelques applications des fonctions elliptiques, *Œuvres de Charles Hermite* **III**, 266-418. (The compilation of the series of articles which appeared in the *Comptes Rendus* from 1877 onwards.)

[1889] Discours d'Hermite dans la séance publique de l'Academie des Sciences en 1889, *Œuvres de Charles Hermite* **IV**, 567-575.

G. W. Hill

[1905] *The Collected Mathematical Works of George William Hill*, 4 volumes, Carnegie Institution of Washington.

[1876] Demonstration of the Differential Equations Employed by Delaunay in the Lunar Theory, *The Analyst* **3**, 65-70 = *Collected Mathematical Works* **I**, No. 26, 227-232.

[1877] On the part of the Motion of the Lunar Perigee which is a Function of the Mean Motions of the Sun and Moon, *John Wilson & Son, Cambridge, Massachusetts*. Reprinted in *Acta* **8**, 1886, 1-36 = *Collected Mathematical Works* **I**, No. 29, 243-270.

[1878] Researches into the Lunar Theory, *American Journal of Mathematics* **I**, 5-26, 129-147, 245-260 = *Collected Mathematical Works* **I**, No. 32, 284-335.

[1896] On the convergence of series used in the subject of perturbations, *Bulletin of the American Mathematical Society* **2**, 93-97 = *Collected Mathematical Works* **IV**, No. 59, 94-98.

[1896a] Remarks on the Progress of Celestial Mechanics since the Middle of the Century, Presidential Address delivered before the American Mathematical Society, 1896, *Bulletin of the American Mathematical Society* **2**, 132-3.

P. Holmes

[1990] Poincaré, celestial mechanics, dynamical-systems theory and "chaos", *Physics Reports* **193** (3), 137-163.

E. Hopf

[1930] Zwei Sätze über den wahrscheinlichen Verlauf der Bewegungen dynamischer System, *Mathematische Annalen* **103**, 710-719.

S. S. Hough

[1901] On certain discontinuities connected with periodic orbits, *Acta* **24**, 257-288 = *G. H. Darwin, Scientific Papers* **IV**, 114-139.

C. G. J. Jacobi

[1836] Sur le mouvement d'un point et sur un cas particulier du problème des trois corps, *Comptes Rendus* **3**, 59-61 = *Gesammelte Werke* **IV**, 35-38.

[1843] Sur l'elimination des noeuds dans le problème des trois corps, *Journal für die Reine und Angewandte Mathematik* **26**, 115-131 = *Gesammelte Werke* **IV**, 295-314.

[1844] Theoria novi multiplicatoris systemati æquationum differentialium vulgarium applicandi, *Journal für due Reine und Angewandte Mathematik* **27**, 199-268; **29**, 213-279, 333-376 = *Gesammelte Werke* **IV**, 317-509.

[1866] *Vorlesungen über dynamik*, Reimer Publisher, Berlin.

Lord Kelvin: see W. Thomson

M. Kline

[1972] *Mathematical Thought from Ancient to Modern Times*, Oxford University Press, New York.

G. Koenigs

[1895] Application des invariants intégraux à la reduction au type canonique d'un système quelques d'équations différentielles, *Comptes Rendus* **121**, 875-878.

A. H. Koblitz

[1983] *A Convergence of Lives. Sophia Kovalevskaia: Scientist, Writer, Revolutionary*, Birkhauser, Boston.

A. N. Kolmogorov

[1954] On conservation of conditionally periodic motions under small perturbations of the Hamiltonian, *Dokl. Akad. Nauk SSSR*,**98**, No. 4, 527-530. (Russian).

S. Kovalevskaya

[1875] Zur theorie der partiellen differentialgleichungen, *Journal für die reine und angewandte Mathematik* **80**, 1-32.

[1978] *A Russian Childhood* (tr. Beatrice Stillman), Springer-Verlag, New York. First edition published in 1889 in Swedish as a novel, *From Russian Life: the Rajevski Sisters*. First published in autobiographical form in Russian in 1890.

L. Kronecker

[1869] Über Systeme von Functionen mehrerer Variabeln, *Monatsberichte der Königlich Preussischen Akademie der Wissenschaften zu Berlin 1869*, 688-698 = *Werke* **I**, 213-226.

[1888] Bemerkungen über Dirichlet's letze Arbeiten, *Sitzungesberichte der Königlich Preussischen Akademie der Wissenschaften 1888*, 439-442 = *Werke* **V**, 471-476.

J. L. Lagrange

[1772] Essai sur le problème des trois corps, *Œuvres* **6**, 229-331.

C. Lanczos

[1970] *The Variational Principles of Mechanics*, 4th edition, University of Toronto Press. Republished by Dover, New York, 1986.

P.-S. Laplace

[1799-1825] *Traité de Mécanique Céleste*, 5 volumes = *Œuvres complètes* **1-5**, Gauthier Villars, Paris, 1891-1904. English translation by Nathanial Bowditch, 1829-1839, reprinted by Chelsea, New York, 1966.

J. La Salle and S. Lefschetz

[1961] *Stability by Liapunov's Direct Method*, Academic Press, New York.

H. Lebesgue

[1902] Intégrale, longeur, aire, *Annali di Matematica Pura ed Applicata* (3) **7**, 231-259.

T. Levi-Civita

[1901] Sopra alcuni criteri di instabilità, *Annali di Mathematica Pura ed Applicata* (3) **5**, 221-307.

[1903] Sur les trajectoires singulières du problème restreint des trois corps, *Comptes Rendus* **135**, 82-84.

[1903a] Condition du choc dans le problème restreint des trois corps, *Comptes Rendus* **135**, 221-223.

[1903b] Traiettorie singolari ed urti nel problema ristretto dei tre corpi, *Annali de Matematica Pura ed Applicata* (3) **9**, 1-32.

[1906] Sur la résolution qualitative du problème restreint des trois corps, *Acta* **30**, 305-327.

[1915] Sulla regolarizzazione del problema piano dei tre corpi, *Rendiconti dell'Accademia Nazionale dei Lincei* (5) **24**, 61-75.

[1918] Sur la régularisation du problème des trois corps, *Acta* **42**, 92-144.

[1926] Sur les chocs dans le problème des trois corps, *Comptes Rendus du 2me Congrès international de mécanique appliquée (Zurich 1926)*, 96-106.

J. Levy

[1952] Notes, *Œuvres de Henri Poincaré* **VII**, 623-632.

A. M. Liapunov

[1907] Problème général de la stabilité du mouvement (tr. A. Davaux), *Annals de la Faculté des Sciences de Toulouse* (2) **9**, 203-474. Reprinted with the same pagination, Princeton University Press, 1947. First edition published in Russian by the Mathematical Society of Kharkow, 1892. Translated from French into English by A. T. Fuller, *International Journal of Control* **55** (3), March 1992, 531-773.

A. Lindstedt

[1883] Beitrag zur integration der differentialgleichungen der störungstheorie, *Mémoires de l'Académie Impériale des Sciences de St. Pétersbourg* (7) **31**, No. 4, 1-19.

[1883a] Sur la forme des expressions des distances mutuelles dans le problème des trois corps, *Comptes Rendus* **97**, 1276-8, 1253-5.

[1884] Sur la détermination des distances mutuelles dans le problème des trois corps, *Annales de l'École Normale* (3) **1**, 85-102.

J. Liouville

[1838] Note sur la théorie de la variation des constantes arbitraires, *Journal de Mathématiques Pures et Appliquées* **3**, 342-349.

E. O. Lovett

[1912] Generalisations of the problem of several bodies, its inversion, and an introductory account of recent progress in its solution, *Quarterly Journal of Pure & Applied Mathematics* **42**, 252-315.

L. Lusternik and L. Schnirrelmann

[1930] Existence de trois géodesiques fermées sur toute surface de genre 0, *Comptes Rendus* **188**, 534-536.

J. Lützen

[1990] *Joseph Liouville 1809-1882*, Springer-Verlag, New York.

W. H. McCrea

[1957] Edmund Taylor Whittaker, *Journal of the London Mathematical Society* **32**, 234-256.

S. W. McCuskey

[1963] *Introduction to Celestial Mechanics*, Addison-Wesley, Reading, Massachusetts.

R. McGehee

[1986] Von Zeipel's theorem on singularities in celestial mechanics, *Expositiones Mathematicae* **4**, 335-345.

R. S. MacKay and J. D. Meiss

[1987] *Hamiltonian Dynamical Systems*, Adam Hilger, Bristol.

W. D. MacMillan

[1913] On Poincaré's correction to Bruns' theorem, *Bulletin of the American Mathematical Society* **15**, 349-355.

J. Mawhin

[1994] The centennial legacy of Poincaré and Lyapunov in ordinary differential equations, *Supplemento ai Rendiconti del Circolo Matematico di Palermo* **34**, Studies in the History of Modern Mathematics I, 9-46.

C. Marchal

[1990] *The Three-Body Problem*, Elsevier, Amsterdam.

R. Marcolongo

[1914] Le recenti ricerche di K. Sundman sul problema matematico dei tre corpi, *Giornale de Matematiche Battaglini* (3) **52**, 171-186.

[1919] *Il problema dei tre corpi da Newton (1686) al nostri giorni*, Hoepli, Milan.

J. Mather and R. McGehee

[1975] Solutions of the collinear four body problem which become unbounded in finite time, *Lecture Notes in Physics* **38**, Springer-Verlag, 573-597.

K R. Meyer and G. R. Hall

[1990] *Introduction to Hamiltonian Dynamical Systems and the N-Body Problem*, Springer-Verlag, New York.

G. Mittag-Leffler

[1902] Une page de la vie de Weierstrass, *Comptes Rendus du deuxième Congrès International des Mathématiciens, tenu a Paris 6-12 aout 1900*, Gauthier Villars, Paris, 131-153.

[1912] Zur Biographie von Weierstrass, *Acta* **38**, 29-65.

M. Morse

[1921] A one-to-one representation of geodesics on a surface of negative curvature, *American Journal of Mathematics* **43**, 33-51.

[1921a] Recurrent geodesics on a surface of negative curvature, *Transactions of the American Mathematical Society* **22**, 84-110.

[1934] *Calculus of variations in the large*, American Mathematical Society Colloquium Publications **18**, Providence, RI.

[1946] George David Birkhoff and his mathematical work, *Bulletin of the American Mathematical Society* **52** (5 i), 357-391.

M. Morse and G. Hedlund

[1938] Symbolic Dynamics, *American Journal of Mathematics* **60**, 815-866.

J. Moser

[1962] On invariant curves of area-preserving mappings of an annulus, *Nachricht von der Akademie der Wissenschaften, Gottingen II, Math. Phys.* **K1**, 1-20.

[1973] *Stable and Random Motions in Dynamical Systems*, Princeton University Press and University of Tokyo Press, Princeton, NJ.

[1978] Is the solar system stable? *Mathematical Intelligencer* **1**, 65-71.

F. R. Moulton

[1914] *An introduction to celestial mechanics*, 2nd edition, Macmillan, New York. Reprinted Dover, New York, 1970.

[1920] *Periodic Orbits* (in collaboration with D. Buchanan, T. Buck, F. L. Griffin, W. R. Longley, W. D. MacMillan), Carnegie Institution of Washington, Washington.

V. V. Nemytskii and V. V. Stepanov

[1960] *Qualitative theory of differential equations*, Princeton University Press, Princeton, NJ.

P. Nabonnand

[1995] Contribution à L'histoire de la théorie des géodésiques ou XIXe sièce, *Revue d'histoire des mathématiques*, **1**, 159–200.

S. Newcomb

[1874] On the General Integrals of Planetary Motion, *Smithsonian Contribution to Knowledge* **31**, 1-31.

Sir Isaac Newton

[1934] *Philosophiæ naturalis principia mathematica* **I - III**, The Royal Society, London, 1687 = *Mathematical Principles of Natural Philosophy,* translated by Andrew Motte, 1729. Revised and edited by Florian Cajori, University of California Press, 1934.

P. Painlevé

[1896] Sur les singularités des équations de la dynamique et sur le problème des trois corps, *Comptes Rendus* **123**, 871-873.

[1897] *Leçons sur la théorie analytique des équations différentielles professées à Stockholm (Septembre, Octobre, Novembre 1895)*, A. Hermann, Paris.

[1897a] Sur les intégrales premières de la dynamique et sur le problème des *n* corps, *Comptes Rendus* **124**, 173-176.

[1897b] Sur le cas du problème des trois corps (et des *n* corps) où deux des corps se choquent au bout d'un temps fini, *Comptes Rendus* **125**, 1078-1081.

[1898] Sur les intégrales premières du problème des *n* corps, *Bulletin Astronomique* **15**, 81-113.

[1900] Sur les intégrales uniformes du problème des *n* corps, *Comptes Rendus* **130**, 1699-1701.

[1912] Henri Poincaré, *Le Temps*, 17 July 1912 = *Acta* **38**, 399-402.

C. Parikh

[1991] *The Unreal Life of Oscar Zariski*, Academic Press, London.

L. A. Pars

[1968] *A Treatise on Analytical Dynamics*, 2nd edition, Heinemann, London.

J. Perchot

[1899] H. Poincaré. Les nouvelles méthodes de la Mécanique céleste, I, *Bulletin des Sciences Mathématiques (2)* **23**, 213-242, 245-260.

E. Phragmén

[1889] Poincaré'ska fallet af tekropparsproblemet (= On some dynamical problems which are related to the restricted three body problem), *K.V.A. Bihang till Handlingar* **15** (1) No. 13, 1-33.

E. Picard

[1896] *Traité d'analyse* **III**, Gauthier Villars, Paris.

[1902] L'Œuvre scientifique de Charles Hermite, *Acta* **25**, 87-111.

[1913] Le problème des trois corps à propos des recherches récentes de M. Sundman, *Bulletin des Sciences mathématiques* **37** (series 2), 313-320.

H. Poincaré

[1915–1956] *Œuvres de Henri Poincaré*, 11 volumes, Gauthier Villars, Paris.

[P1] Sur le problème des trois corps et les équations de la dynamique avec des notes par l'auteur - mémoire couronné du prix de S. M. le Roi Oscar II. Printed in 1889 but not published.

[P1a] As [P1] but with handwritten corrections and additions by the author.

[P2] Sur le problème des trois corps et les équations de la dynamique, *Acta* **13**, 1890, 1-270 = *Œuvres* **VII**, 262-479.

[MN I] *Les Méthodes Nouvelles de la Mécanique Céleste* **I** , Gauthier-Villars, 1892. Republished by Blanchard, Paris, 1987.

[MN II] *Les Méthodes Nouvelles de la Mécanique Céleste* **II** , Gauthier-Villars, 1893. Republished by Blanchard, Paris, 1987.

[MN III] *Les Méthodes Nouvelles de la Mécanique Céleste* **III** , Gauthier-Villars, 1899. Republished by Blanchard, Paris, 1987.

[LMC] *Leçons de Mécanique Céleste* **I-III**, Gauthier-Villars, 1905-1910.

[1879] Sur les propriétés des fonctions définies par les équations aux différence partielles, Première Thèse, Gauthier-Villars = *Œuvres* **I**, XLIX –CXXIX.

[1880] Sur les courbes définies par une équation différentielle, *Comptes Rendus* **90**, 673-675 = *Œuvres* **I**, 1-2.

[1881] Mémoire sur les courbes définies par une équation différentielle, *Journal de Mathématiques* (3) **7**, 375-422 = *Œuvres* **I**, 3-44.

[1882] Mémoire sur les courbes définies par une équation différentielle, *Journal de Mathématiques* (3) **8**, 251-296 = *Œuvres* **I**, 44-84.

[1882a] Sur l'intégration des équations différentielles par les series, *Comptes Rendus* **94**, 577-578 = *Œuvres* **I**, 162-163.

[1882b] Sur les séries trigonométriques, *Comptes Rendus* **95**, 766-768 = *Œuvres* **IV**, 585-587.

[1883] Sur certaines solutions particulières du problème des trois corps, *Comptes Rendus* **97**, 251-252 = *Œuvres* **VII**, 251-252.

[1883a] Sur les séries trigonométriques, *Comptes Rendus* **97**, 1471-1473 = *Œuvres* **IV**, 588-590.

[1884] Notice sur les traveaux scientifiques, *Gauthier-Villars*.

[1884a] Sur certaines solutions particulières du problème des trois corps, *Bulletin Astronomique* **1**, 65-74 = *Œuvres* **VII**, 253-261.

[1884b] Sur la convergence des séries trigonométriques, *Bulletin Astronomique* **1**, 319-327 = *Œuvres* **IV**, 591-598.

[1884c] Sur une équation différentielle, *Comptes Rendus* **98**, 793-5 = *Œuvres* **VII**, 543-545.

[1885] Sur les courbes définies par les équations différentielles, *Journal de Mathématiques* (4) **1**, 167-244 = *Œuvres* **I**, 90-161.

[1885a] Sur les séries trigonométriques, *Comptes Rendus* **101**, 1131-1134 = *Œuvres* **I**, 164-166.

[1885b] Sur l'équilibre d'une masse fluide animée d'un mouvement de rotation, *Acta* **7**, 259-380 = *Œuvres* **VII**, 40-140.

[1886] Sur les courbes définies par les équations différentielles, *Journal de Mathématiques* (4) **2**, 151-217 = *Œuvres* **I**, 167-222.

[1886a] Sur les intégrales irrégulières des équations linéaires, *Acta* **8**, 295-344 = *Œuvres* **I**, 290-332.

[1886b] Sur une méthode de M. Lindstedt, *Bulletin Astronomique* **3**, 57-61 = *Œuvres* **VII**, 546-550.

[1886c] Sur un moyen d'augmenter la convergence des séries trigonométriques, *Bulletin Astronomique* **3**, 521-8 = *Œuvres* **IV**, 599-606.

[1886d] Sur les déterminants d'ordre infini, *Bulletin de la Société mathématique de France* **14**, 77-90 = *Œuvres* **V**, 95-107.

[1889] Sur les séries de M. Lindstedt, *Comptes Rendus* **108**, 21-24 = *Œuvres* **VII**, 551-554.

[1891] Sur le problème des trois corps, *Bulletin Astronomique* **8**, 12-24 = *Œuvres* **VII**, 480-490.

[1891a] Le problème des trois corps, *Revue Générale des Sciences pures et appliquées* **2**, 1-5 = *Œuvres* **VIII**, 529-537.

[1891b] Sur le développement approché de la fonction perturbatrice, *Comptes Rendus* **112**, 269-273 = *Œuvres* **VIII**, 5-9.

[1892] Sur l'application de la méthode de M. Lindstedt au problème des trois corps, *Comptes Rendus* **114**, 1305-1309 = *Œuvres* **VII**, 491-495.

[1896] Sur la divergence des séries de la mécanique céleste, *Comptes Rendus* **122**, 497-499 = *Œuvres* **VII**, 558-560.

[1896a] Sur la divergence des séries trigonométriques, *Comptes Rendus* **122**, 557-559 = *Œuvres* **VII**, 561-563.

[1896b] Sur la méthode de Bruns, *Comptes Rendus* **123**, 1224-1228 = *Œuvres* **VII**, 512-516.

[1896c] Sur les solutions périodiques et le principe de moindre action, *Comptes Rendus* **123**, 915-918 = *Œuvres* **VII**, 224-226.

[1897] Sur les solutions périodiques et le principe de moindre action, *Comptes Rendus* **124**, 713-716 = *Œuvres* **VII**, 227-230.

[1898] On the stability of the solar system, *Nature* **58**, 183-184. Originally published in *Annuaire du Bureau des Longitudes* and *Revue Scientifique* **9** (4), 609-613 = *Œuvres* **VIII**, 538-547.

[1901] Sur la théorie de précession, *Comptes Rendus* **132**, 50-55 = *Œuvres* **VIII**, 113-117.

[1904] Sur la méthode horistique de Gyldén, *Comptes Rendus* **138**, 933-936 = *Œuvres* **VII**, 583-586.

[1904a] Sur la méthode horistique. Observations sur l'article de M. Backlund, *Bulletin Astronomique* **21**, 292-295 = *Œuvres* **VII**, 619-621.

[1905] *Introduction to The Collected Mathematical Works of George William Hill, Volume 1*, Carnegie Institution of Washington, VII-XVIII.

[1905a] Sur la méthode horistique de Gyldén, *Acta* **29**, 235-271 = *Œuvres* **VII**, 587-618.

[1905b] Sur les lignes géodésiques des surfaces convexes, *Transactions of the American Mathematical Society* **6**, 237-274 = *Œuvres* **VI**, 38-84.

[1912] Sur un théorème de géométrie, *Rendiconti Circolo matematico di Palermo* **33**, 375-407 = *Œuvres* **VI**, 499-538.

A. Roy

[1968] *Orbital Motion*, Adam Hilger, Bristol.

D. Saari

[1990] A visit to the Newtonian *n*-body problem via elementary complex variables, *The American Mathematical Monthly* **97**, 105-119.

G. Sarton

[1913] Henri Poincaré, *Ciel et Terre*. *Bulletin de la Société belge d'Astronomie*, 1-11, 37-48.

A. Schlissel

[1976] The development of asymptotic solutions of linear ordinary differential equations 1817-1920, *Archive for History of Exact Sciences* **16**, 307-378.

E. Scholz

[1980] *Geschichte des Mannigfaltigkeitsbegriffs von Riemann bis Poincaré*, Birkhauser, Boston and Basel.

K. Schwarzschild

[1898] Über eine Classe periodischer Lösungen des Dreikörperproblems, *Astronomische Nachrichten* **147**, 17, 289.

C. L. Siegel and J. K. Moser

[1971] *Lectures on Celestial Mechanics*, Springer-Verlag, Berlin. First edition published as *Vorlesungen über Himmelsmechanik*, Springer, 1956.

V. I. Smirnov

[1992] Biography of A. M. Liapunov, *International Journal of Control* **55** (3), 775-784. Translated by J. F. Barrett from the Russian article in *A. M. Lyapunov: Izbrannie Trudie* (A. M. Liapunov: Selected Works) (Leningrad: Izdat Akad. Nauk SSR, 1948), edited by V. I. Smirnov, 325-340.

D. E. Smith and J. Ginsburg

[1934] *A History of Mathematics in America before 1900*, The Mathematical Association of America (Carus Mathematical Monograph Number 5).

S. Sternberg

[1969] *Celestial Mechanics, Parts I and II*, W. A. Benjamin, New York.

I. Stewart

[1989] *Does God Play Dice ?*, Blackwell, Oxford.

K. F. Sundman

[1907] Recherches sur le problème des trois corps, *Acta Societatis Scientiarum Fennicae* **34**, No. 6, 1-43.

[1909] Nouvelles recherches sur le problème des trois corps, *Acta Societatis Scientiarum Fennicae* **35**, No. 9, 1-27.

[1912] Mémoire sur le probléme des trois corps, *Acta* **36**, 105-179.

V. Szebehely

[1967] *Theory of Orbits*, Academic Press, New York and London.

W. Thomson (Lord Kelvin)

[1891] On periodic motion of a finite conservative system, *Philosophical Magazine* **32**, 375-383; Instability of Periodic Motion, *Philosophical Magazine* **32**, 555-560 = *Mathematical and Physics Papers*, **IV**, Cambridge University Press, 1910, 497-512.

F. Tisserand

[1887] Sur la commensurabilité des moyens mouvements dans le système solaire, *Comptes Rendus* **104**, 259-265; *Bulletin Astronomique* **4**, 183-192.

[1889-1896] *Traité de Mécanique Céleste* **I**, 1889; **II**, 1891; **III**, 1894; **IV**, 1896; Gauthier-Villars, Paris.

O. Veblen

[1946] George David Birkhoff, *Yearbook American Philosophical Society*, 279-285.

E. B. Van Vleck

[1915] The role of the point-set theory in geometry and dynamics, *Bulletin of the American Mathematical Society* **21**, 321-341.

H. von Zeipel

[1908] Sur les singularities du problème des *n* corps, *Arkiv för Matematik, Astronomi och Fysik* **4** (32), 1-4.

[1921] L'Œuvre Astronomique de Henri Poincaré, *Acta* **38**, 309-385.

R. S. Westall

[1980] *Never at Rest: A Biography of Isaac Newton*, Cambridge University Press.

E. T. Whittaker

[1899] Report on the progress of the solution of the problem of three bodies, *BAAS Report 1899*, 121-159.

[1901] On the solution of dynamical problems in terms of trigonometric series, *Proceedings of the London Mathematical Society* **34**, 206-221.

[1902] Periodic orbits, *Monthly Notices of the Royal Astronomical Society* **62**, 186-193.

[1902a] On periodic orbits in the restricted problem of three bodies, *Monthly Notices of the Royal Astronomical Society* **62**, 346-352.

[1912] Prinzipien der Störungstheorie und allgemeine Theorie der Bahnkurven in dynamischen Problemen, *Encyklopädie der mathematische Wissenschaften* **VI** (2), 512-556.

[1917] On the adelphic integral of the differential equations of dynamics, *Proceedings of the Royal Society of Edinburgh* **37**, 95-116.

[1937] *A treatise on the analytical dynamics of particles and rigid bodies*, 4th edition, Cambridge University Press. First edition published 1904.

E. T. Whittaker and G. N. Watson

[1927] *A course of modern analysis*, 4th edition, Cambridge University Press. First edition published 1902.

S. Wiggins

[1993] *Global Dynamics, Phase Space Transport, Orbits Homoclinic to Resonances and Applications*, American Mathematical Society, Providence, RI.

A. Wintner

[1941] *The Analytical Foundations of Celestial Mechanics*, Princeton University Press, Princeton, NJ.

Z. Xia

[1992] The existence of noncollision singularities in Newtonian systems, *Annals of Mathematics* **135**, 411-468.

Index